Lecture Notes in Artificial Intelligence   4303

Edited by J. G. Carbonell and J. Siekmann

Subseries of Lecture Notes in Computer Science

Achim Hoffmann   Byeong-ho Kang
Debbie Richards   Shusaku Tsumoto (Eds.)

Advances in Knowledge Acquisition and Management

Pacific Rim Knowledge Acquisition Workshop, PKAW 2006
Guilin, China, August 7-8, 2006
Revised Selected Papers

 Springer

Series Editors

Jaime G. Carbonell, Carnegie Mellon University, Pittsburgh, PA, USA
Jörg Siekmann, University of Saarland, Saarbrücken, Germany

Volume Editors

Achim Hoffmann
The University of New South Wales, School of Computer Science & Engineering
Sydney 2052, Australia
E-mail: achim@cse.unsw.edu.au

Byeong-ho Kang
University of Tasmania, School of Computing
Hobart Campus, Centenary Building, Hobart, TAS 7001, Australia
E-mail: bhkang@utas.edu.au

Debbie Richards
Macquarie University, Department of Computing
Sydney, NSW 2109, Australia
E-mail: richards@ics.mq.edu.au

Shusaku Tsumoto
Shimane University, School of Medicine, Department of Medical Informatics
89-1 Enya-cho, Izumo 693-8501, Japan
E-mail: tsumoto@computer.org

Library of Congress Control Number: 2006938401

CR Subject Classification (1998): I.2.6, I.2, H.2.8, H.3-5, F.2.2, C.2.4, K.3

LNCS Sublibrary: SL 7 – Artificial Intelligence

ISSN   0302-9743
ISBN-10   3-540-68955-9 Springer Berlin Heidelberg New York
ISBN-13   978-3-540-68955-3 Springer Berlin Heidelberg New York

Springer is a part of Springer Science+Business Media

springer.com

© Springer-Verlag Berlin Heidelberg 2006
Printed in Germany

Typesetting: Camera-ready by author, data conversion by Scientific Publishing Services, Chennai, India
Printed on acid-free paper   SPIN: 11961239   06/3142   5 4 3 2 1 0

Preface

Since knowledge was recognized as a crucial part of intelligent systems in the 1970s and early 1980s, the problem of the systematic and efficient acquisition of knowledge was an important research problem. In the early days of expert systems, the focus of knowledge acquisition was to design a suitable knowledge base for the problem domain by eliciting the knowledge from available experts before the system was completed and deployed. Over the years, alternative approaches were developed, such as incremental approaches which would build a provisional knowledge base initially and would improve the knowledge base while the system was used in practice. Other approaches sought to build knowledge bases fully automatically by employing machine-learning methods. In recent years, a significant interest developed regarding the problem of constructing ontologies. Of particular interest have been ontologies that could be re-used in a number of ways and could possibly be shared across different users as well as domains.

The Pacific Knowledge Acquisition Workshops (PKAW) have a long tradition in providing a forum for researchers to exchange the latest ideas on the topic. Participants come from all over the world but with a focus on the Pacific Rim region. PKAW is one of three international knowledge acquisition workshop series held in the Pacific-Rim, Canada and Europe over the last two decades. The previous Pacific Knowledge Acquisition Workshop, PKAW 2004, had a strong emphasis on incremental knowledge acquisition, machine learning, neural networks and data mining.

This volume contains the post-proceedings of the Pacific Knowledge Acquisition Workshop 2006 (PKAW 2006) held in Guilin, China. The workshop received 81 submissions from 12 countries. All papers were refereed in full length by the members of the International Program Committee. A very rigorous selection process resulted in the acceptance of only 21 long papers (26%) and 6 short papers (7.5%). Revised versions of these papers which took the discussions at the workshop into account are included in this post-workshop volume. The selected papers show how the latest international research made progress in the above-mentioned aspects of knowledge acquisition. A number of papers also demonstrate practical applications of developed techniques.

The success of a workshop depends on the support of all the people involved. Therefore, the workshop Co-chairs would like to thank all the people who contributed to the success of PKAW 2006. First of all, we would like to take this opportunity to thank authors and participants. We wish to thank the Program Committee members who reviewed the papers and the volunteer student Yangsok Kim at The University of Tasmania for the administration of the workshop.

August 2006   Achmin Hoffmann
                                                                                             Byeoug-ho Kang
                                                                                             Debbie Richards
                                                                                             Shusaku Tsumoto

Organization

Honorary Chairs

Paul Compton (University of New South Wales, Australia)
Hiroshi Motoda (Osaka University, Japan)

Workshop Co-chairs

Achim Hoffmann (University of New South Wales, Australia)
Byeong-ho Kang (Tasmania University, Australia)
Debbie Richards (Macquarie University, Australia)
Shusaku Tsumoto (Shimane University, Japan)

Program Committee

George Macleod Coghill (University of Aberdeen, UK)
Rob Colomb (University of Queensland, Australia)
John Debenham (University of Technology, Sydney, Australia)
Rose Dieng (INRIA, France)
Fabrice Guillet (L'Université de Nantes, France)
Udo Hahn (Freiburg University, Germany)
Ray Hashemi (Armstrong Atlantic State University, USA)
Noriaki Izumi (Cyber Assist Research Center, AIST, Japan)
Yasuhiko Kitamura (Kwansei Gakuin University, Japan)
Mihye Kim (Catholic University of Daegu, Korea)
Rob Kremer (University of Calgary, Canada)
Huan Liu (Arizona State University, USA)
Ashesh Jayantbhai Mahidadia (University of New South Wales, Australia)
Stephen MacDonell (Auckland Universtiy of Techonology, New Zealand)
Rodrigo Martinez-Bejar (University of Murcia, Spain)
Tim Menzies (NASA, USA)
Kyongho Min (Auckland University of Technology, New Zealand)
Toshiro Minami (Kyushu Institute of Information Sciences and Kyushu University, Japan)
Masayuki Numao (Osaka University, Japan)
Takashi Okada (Kwansei Gakuin University, Japan)
Frank Puppe (University of Wuerzburg, Germany)
Ulrich Reimer (Business Operation Systems, Switzerland)
Debbie Richards (Macquarie University, Australia)
Masashi Shimbo (Nara Institute of Science and Technology, Japan)
Hendra Suryanto (University of New South Wales, Australia)

Takao Terano (University of Tsukuba, Japan)
Peter Vamplew (University of Ballarat, Australia)
Takashi Washio (Osaka University, Japan)
Ray Williams (University of Tasmania, Australia)
Shuxiang Xu (University of Tasmania, Australia)
Seiji Yamada (National Institute of Informatics, Japan)

Table of Contents

Incremental Knowledge Acquisition and RDR

Knowledge Acquisition and Applications

Machine Learning and Data Mining

PART II: Regular Papers (6–8 Pages)

Visual Knowledge Annotation and Management by Using Qualitative Spatial Information

Pedro José Vivancos-Vicente, Jesualdo Tomás Fernández-Breis,
Rodrigo Martínez-Béjar, and Rafael Valencia-García

Tecnologías del Conocimiento y Modelado Cognitivo (TECNOMOD) Group
Facultad de Informática, Campus de Espinardo. 30071 Murcia. Spain
{pedroviv, jfernand, rodrigo, valencia}@um.es
http://klt.inf.um.es

Abstract. The wide use of the Internet and the increasingly improvement of communication technologies have led users to need to manage multimedia information. In particular, there is an ample consensus about the necessity of new computational systems capable of processing images and "understand" what they contain. Such systems would ideally allow to retrieve multimedia content, to improve the way of storing it or to process the images to get some information interesting for the user. This paper presents a methodology for semi-automatically extracting knowledge from 2D still visual multimedia content, that is, images. The knowledge is acquired through the combination of several approaches: computer vision (to get and to analyse low level features), qualitative spatial analysis (to obtain high level information from low level features), ontologies (to represent knowledge), and MPEG-7 (to describe the information in a standard-way and make the system capable of performing queries and retrieve multimedia content).

1 Introduction

An incommensurable amount of visual information is becoming available in digital form, in digital archives, on the World Wide Web, in broadcast data streams and in personal and professional databases, and this kind of information is increasing.

Nowadays, it is common to have access to powerful computers capable of executing complex processes and despite that, there is no efficient approach to process multimedia content to extract high-level features (as knowledge) from them. Moreover, a lot of processes use multimedia contents as their primary data source in critical domains.

It is clear new computational systems capable of processing and "understanding" multimedia content are needed. So, different processes can be performed more efficiently: multimedia content retrieval, storage and processing. In this way, images can be processed to get interesting information for the user, who is not interested in low-level features of multimedia information but in high-level ones (i.e., the content meaning). This is the so-called semantic gap: how to bridge the low-level features and high-level features. It refers to the cognitive distance between the analysis results delivered by state-of-art image-analysis tools and the concepts human look for in images [4].

A. Hoffmann et al. (Eds.): PKAW 2006, LNAI 4303, pp. 1–12, 2006.

Traditionally, textual features such as filenames, captions, and keywords have been used to annotate and retrieve images [7]. Research on intelligent systems for extracting knowledge or meta-information directly from multimedia content has increased in the last years. For example, systems which usually work with sport videos, recognising some kinds of events as a function of audio comments [12,13,14]. But this is not enough to get meta information about the image content. Many content-based image retrieval systems have been proposed in literature [1,2,3]. Most of them try to get more information by analysing the image to work out low-level features such as colours, textures, and shapes of objects, but this is not sufficient to get real information about what an image contents [11].

In this work, an approach to obtain high-level features from images using ontologies and qualitative spatial representation and reasoning is presented. This approach extracts relationships between the regions of the image by using their low-level features obtained in the segmentation step. Then, it creates a content representation where the regions are concepts, the low-level features their attributes and the relationships are inferred knowledge. This information is then used to compare this structure to ontologies stored in libraries so that the system can guess what each region really is and perhaps, what the image represents. An advantage of using semantic approaches is the fact that they do not require to re-design the framework for different domains. It provides a new layer that is completely independent of the methods and techniques used to process the image.

Finally, the structure of this paper is the following. In section 2, the technical background of this methodology is discussed. An overview of the methodology proposed for this work is described in section 3. Section 4 describes the processes for extracting high-level information from images. An example of the methodology is shown in section 5. Finally, some conclusions are put forward in section 6.

2 Technical Background

Along this section the basic methodological components of our approach are briefly explained.

2.1 Image Segmentation

Image segmentation is a challenging and important issue in image processing and computer vision. It tries to extract the objects an observer may find conceptually coherent by themselves, so that the extracted objects (i.e., regions) are distinct from each other. However, segmentation has access only to the descriptions of pixels (i.e., colour), and their spatial relationships, while a human observer always uses a higher level of knowledge (e.g., object recognition) to segment the image objects.

There are many segmentation algorithms, which are usually specialized in extracting specific types of regions (i.e., the background) [16]. Moreover, some can be used together to get different kind of information and then try to merge it.

To us, segmentation will provide a set of regions that will be used to get high-level information. So, after segmentation, an image is decomposed into a set of regions for which the system must try to find out their real meaning. Once segmentation has been performed, a set of low-level features are obtained for each region.

2.2 Qualitative Spatial Reasoning

According to [5], Qualitative Reasoning deals with capturing the knowledge of physical world while creating quantitative models. The ultimate objective of Qualitative Reasoning is to make this knowledge explicit, so that from appropriate reasoning techniques, a computer might predict, diagnose and explain the behaviour of physical systems in a qualitative manner.

Qualitative spatial representations address many different aspects of space including topology, orientation, shape, size and distance, so it has already been used in computer vision for visual object recognition at a higher level, including the interpretation and integration of visual information. It has been used to interpret the results of low-level computations as higher level descriptions of the scene or video input [10]. The use of qualitative predicates helps to ensure that semantically similar scenes have identical or at least very similar descriptions.

2.3 Ontologies

An ontology is viewed in this work as a formal specification of a domain knowledge conceptualization [15]. In this sense, ontologies provide a formal, structured knowledge representation, having the advantage of being reusable and shareable. Furthermore, an ontology can be seen as a semantic model containing concepts, their properties, interconceptual relations, and axioms related to the previous elements. For our purpose, ontologies can represent topological information of different domains, so this knowledge is used to infer the meaning of an image. This usage is discussed in [5], where the authors state that there are strong reasons for taking regions as the ontological primitive.

2.4 MPEG-7

MPEG-7, formally named "Multimedia Content Description Interface", is a ISO/IEC standard developed by MPEG for describing multimedia content data that supports some degree of interpretation of the information meaning, which can be passed onto, or accessed by, a device or a computer code [8].

MPEG-7 offers a comprehensive set of audiovisual Description Tools (the metadata elements and their structure and relationships, that are defined by the standard in the form of Descriptors and Description Schemes) to create descriptions. These descriptions are a set of instantiated Description Schemes and their corresponding Descriptors at the users will form the basis for applications enabling the needed effective and efficient access (search, filtering and browsing) to multimedia content. This is a challenging task given the broad spectrum of requirements and targeted multimedia applications, and the broad number of audiovisual features of importance in such context.

3 Overview of the Methodology

The methodology proposed here can be visualised through a framework comprised of three main modules, namely, image processing and low level features extraction, qualitative spatial information extraction and ontology library sub-system.

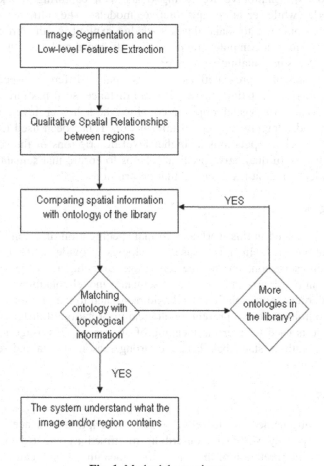

Fig. 1. Methodology schema

The goal of the framework is to get high level information from an input image. This process may be supported by an expert during the image processing task to extract the image segments.

At the beginning, some filters and techniques are applied to the input image to determine the segments. This step may be done by an expert if the content of the image is not previously known. This may be done (semi) automatically if the system has analysed similar images (the same domain) before, so it knows which algorithms and techniques to apply. After that, the system knows every segment that composes the whole image; so, it is possible to extract low level features for each segment and for the whole image in general (e.g. the background dominant colour). Once segmentation has been performed, a set of low-level features are obtained for each segment. This process will be explained in the next section.

Another sub-system will be able to get qualitative spatial information between segments so a structure of topological relationships between concepts will be obtained (e.g. "A is left of B", "C is similar size to D"...)

Once the structure is obtained, it can be compared with all the ontologies with topological information stored in our library subsystem to match the structure with an ontology. In this case, the system is able to interpret the image in the context of a particular domain.

4   High-Level Information Extraction Process

After describing the framework, the focus of this section is the knowledge extraction processes carried out. For each image, the system performs three sequential phases: image segmentation and low level features extraction, qualitative spatial relationships extraction and inference.

4.1   Image Processing and Low Level Features Extraction

As it has been abovementioned, the input is an image. The system processes each image to obtain the elements that appear in the image, using several techniques based on segmentation, which is the process that partitions the spatial domain of an image or other raster datasets like digital elevation models into mutually exclusive parts, called regions. After that, the system gets several quantitative features [6] for each element found. Theses features are enumerated in the following list:

Table 1. Features description

Features Description	
Position	It is defined as the portion of space that is occupied by the object.
Orientation	Where the object is pointing to.
Location	refers to the location of the object in the image (e.g. far north)
Size	The area of the segment.
Compactness	Represents the density of an object.
Dimension	It is composed by two properties: width and height.
Perimeter	It represents the distance around a figure.
Shape	It represents the visual appearance of a region.
Colour	The visual attribute of the region that results from the light they emit or transmit or reflect.
Texture	The tactile quality of a surface or the representation or invention of the appearance of such a surface quality.

Once all this information has been obtained, it may be represented in MPEG-7 because it has descriptors to represent general information about the image and for each region. Some of them are to define basic structures like colour, texture, shape, localization and others for another kind of multimedia contents such us video or sounds.

4.2 Qualitative Spatial Information Extraction

Qualitative representation has already been used in computer vision for visual object recognition at a higher level, which includes the interpretation and integration of visual information. The use of qualitative spatial information helps to ensure that semantically close scenes have highly similar descriptions. Hence, it is possible to recognise images that represent the same content. Our approach uses ontologies to define topologically a scenario. So, the system can compare the information obtained from the image with the ontologies in order to infer the content of an image. So, in this way, the system is able to interpret the results of low-level computations as higher level scene descriptions.

In order to achieve our objective, the system must find out all the spatial relationships existing between all the regions detected in the image through the previous phase. These relationships will give us information about how the regions are spatially 'related', that is, how the scenario is configured. The result of this process is a 'graph' where the nodes or concepts (regions) are related to each other by using different kinds of spatial relationships. The spatial relationships the system can work with are explained below.

4.2.1 Location Relationships

The location of one object is relative to another object. The first one is called the 'target' object and the second is the 'reference' object. Generally, this location has two components, a distance and a direction, so the system gets relations such as "A is north of B" and "A is far from B".

The system is able to work out the following location (cardinal) relationships: east, west, north, south, south-west, north-west, south-east, north-east and a special case called central area, in which other relations between two objects cannot be applied [9]. On the other hand, the system can also calculate some relations related to distance between objects: near, mid and far which may be obtained using fuzzy logic [10].

4.2.2 Orientation Relationships

The system can handle one orientation relationship, "pointing to". This relationship is not easy to get in an isolated image. Sometimes it is necessary to notice the movement of an element in a sequence of images to determine its orientation [6].

4.2.3 Dimension Relationships

By using this attribute, the system is able to get relationships about the height and width differences between objects, such as "A is similar height to B".

4.2.4 Size Relationships

In contrast to dimension relationships, where the system may obtain relations of height and width, size relationships compare the area of the elements (the system only can only work with 2D images). So, comparing objects using the 'size' property makes the system get relationships such as 'X is similar size to Y', where X and Y are elements of the image.

4.2.5 Connectivity Relationships

Connectivity relationships define the degree of connectedness of two elements of the image. The topological definition of connectedness is that a point-set is connected if, and only if, whenever it is contained in the union of two disjoint open sets, it is wholly contained in just one of those sets [6].

The system is capable of working with the following relations: part of, overlap, and partially overlap. An example of these relationships could be: "Iris overlaps pupil", in an eye.

4.2.6 Shape Relationships

Shape is a complex property of the image elements. The spatial relationships between two regions of the image represent that some objects have a similar shape.

4.2.7 Colour Relationships

Although colour is not a spatial or a geometric property, it is interesting to find some similarities between objects in an image. It is possible to use fuzzy logic to say that an object (X) has similar colour to another (Y), trying to get similarities between them.

As it has been described before, once all the spatial information has been obtained, it is possible to represent all this information in MPEG-7. Although there are no descriptors to represent spatial relationships, new descriptors may be created to store the information obtained in MPEG-7, so that higher level information of the image is also stored in the same description.

4.3 Ontology Library Sub-system

Let us suppose that the system has obtained concept instances of an ontology with topological information. In this case, it would be possible to compare this ontology to all the ontologies of a library with the purpose of matching one of these ones with the concept instances the system has (checking if all the axioms are fulfilled). In this way, the system would be capable of inferring what an image contains.

Let us suppose an image of a human face as the system input. The system would be able to get some information about the concepts (elements) that appear in the image (e.g. "A has a similar shape to B"). However, the system does not know what each element is. So, the system compares the information obtained with the structure and axioms of an ontology, trying to infer whether the elements of the image are instances of the ontology concepts. In this case, the system could interpret the element and, eventually, the whole image (or part of it) with respect to that domain ontology. The result of this last process is real high-level information of the image, i.e. the human interpretation of the image, that is, information interesting for a human being.

As it has been pointed out before, once it is known what an image contains it can also be stored using MPEG-7 descriptions, so that high-level information is available for future searches.

5 Example

Let us illustrate how the methodology works through a very simple example. Let us suppose that the system must analyse the following image of a head. It should be noted that the system does not know what it is.

Fig. 2. Sample image

In this case, a human being can easily see a mouth, two eyes, a nose and two ears, that is, a head. Let us see step by step how the system might come to the same conclusion.

Step 1: Image segmentation

As it has been described before, the segmentation process is a difficult task. It usually needs the support of an expert (at least once for each kind of image/domain) to get all the regions. In this example, the image to process is already segmented (notice that each region is of a colour different from the around regions). So, the result of the segmentation is shown in figure 3:

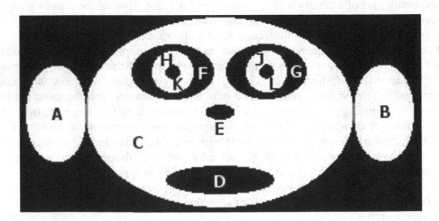

Fig. 3. Labeled image

Each region has been labeled so a human being would say that A and B are ears, C is the whole face, D is the mouth, E is the nose, and F and G are the eyes.

For each region, the system obtains all the (qualitative or quantitative) attributes we mentioned before: position, size, etc so that the next step can be performed.

Step 2: Qualitative Spatial Information Extraction

Once the segmentation task has been performed, the system uses the information obtained in the previous step to get qualitative spatial relationships between all the regions of the image. In our example, some of the relationships the system may get are shown in the following table:

Table 2. Some spatial relationships obtained for the example image

LEFT(A,B)	SIM_SIZE(A,B)	ABOVE(E,D)	SIM_SHAPE(A,B)	PART_OF(C,D)
LEFT(A,C)	SIM_SIZE(F,G)	ABOVE(A,D)	SIM_SHAPE(F,G)	PART_OF(C,E)
LEFT(A,D)		ABOVE(B,D)		PART_OF(C,F)
LEFT(A,E)		ABOVE(F,D)		PART_OF(C,G)
LEFT(A,F)		ABOVE(G,D)		
LEFT(A,G)		ABOVE(F,E)		

Where SIM_SIZE(x,y) means "x has a similar size to y" and SIM_SHAPE(x,y) means "x has a similar shape to y", both are symmetric relationships. LEFT and ABOVE functions have inverse relationships (i.e., RIGHT and BELOW, respectively).

Step 3: Comparing the structure obtained with the ontologies in the library

Once the system has found all the regions and the relationships between them, it will be capable of comparing the structure obtained with the ontologies of the library in order to determine whether the image represents something described in an ontology.

In our knowledge base, a head is described by using the ontology shown below.

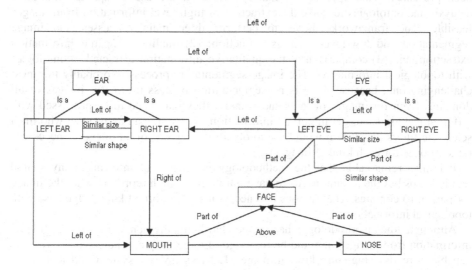

Fig. 4. Ontology of a "head"

Notice that some relationships have been omitted, such as "right of", because we have considered "left of" (corresponding to the symmetric relationship of "right of").

Our system will be capable of comparing the ontology and the spatial information obtained to guess that the image contains a head, and what it was labeled as A, B, C,... are LEFT EAR, RIGHT EAR, FACE, and so on, respectively. The matching is based on the ontology structural axioms. Each concept has a list of structural axioms (e.g., "A is a concept", "A is related to B",...) and each rule obtained is a potential axiom. If a subset of rules characteristing an object in the image accomplish with all the axioms of a concept, the system infers this object is an instance of such a concept.

LEFT(A,B)	LEFT(LEFT_EAR,RIGHT_EAR)	
LEFT(A,C)	LEFT(LEFT_EAR,FACE)	
LEFT(A,D)	LEFT(LEFT_EAR,MOUTH)	
LEFT(A,E)	LEFT(LEFT_EAR,NOSE)	
LEFT(A,F)	LEFT(LEFT_EAR,LEFT_EYE)	
LEFT(A,G)	LEFT(LEFT_EAR,RIGHT_EYE)	$\Rightarrow$ A is a LEFT_EAR
SIM_SIZE(A,B)	SIM_SIZE(LEFT_EAR, RIGHT_EAR)	
ABOVE(A,D)	ABOVE(LEFT_EAR, MOUTH)	
SIM_SHAPE(A,B)	SIM_SHAPE(LEFT_EAR, RIGHT_EAR)	
....	....	

Fig. 5. Matching of concepts

6   Discussion and Conclusions

In this paper, a semi-automatic method for acquiring knowledge from images has been presented. This approach is mainly based on the use of spatial qualitative analysis and ontologies to make the extraction of high-level information from images feasible. The framework allows to perform three main processes: (1) image segmentation and low-level features extraction; (2) qualitative spatial information extraction; and (3) comparison of the qualitative information extracted to ontologies with topological information. The image segmentation process is probably the most challenging one, because there is no segmentation process useful for images of all domains. Even for images of the same domain, they may need to be processed with different techniques to get important information. So the human assistance, who must select the filters the system would use to obtain the regions of an image, is necessary (at least once for each kind of images).

It is also remarkable that the methodology is capable of inferring many spatial relationships but the system needs to have ontologies which represent what the image (or part of it) contains. That implies it is necessary an important knowledge base with topological information.

Although the methodology has some hard requirements to get high-level information from images, it could be really good to develop computer vision systems capable of recognising some kind of images. Let us suppose a system to detect breast tumors is to be implemented. All the images will be quite similar to each other, so the computer vision system developer just needs to define the filters to be applied in the segmentation process. After that, a knowledge engineer must define an ontology for the domain under question with topological information, so that the system will be able to detect breast lumps and even say if it is malign or not (by using the information in the ontology).

We are currently developing a software system which will implement this methodology and will be used in a medical domain to detect some kind of tumors semi automatically. To be more precise, the system will detect automatically every part of the body from the image and infer if there is something wrong, detecting quickly a possible tumor. This high-level information will be stored in MPEG-7 format in order to make the information available easily for the hospital staff.

Acknowledgements

This work has been possible thanks to the Spanish Ministry for Science and Education through the projects TSI2004-06475-C02; the Seneca Foundation through the Project 00670/PI/04; FUNDESOCO through project FDS-2004-001-01; the Autonomous Community of Murcia Region through project 2I05SU0013; the European Commission ALFA through project FA-0447.

References

[1] Flickner, M., Shawney, H., Niblack, W., Ashley, J., Huang, Q., Dom, B., Query by image and video content: the QBIC system. IEEE Computer, 28(9), 23-32. 1995

[2] Jain, A.K., Vailaya, K. Image retrieval using color and shape. Pattern recognition, 29, 1233-1244. 1996

[3] Rui, Y., Huang, T.S., Chang, S.F. Image retrieval: current techniques, promising directions, and open issues. Journal of Visual Communications and Image representation, 10, 39-62. 1999

[4] Hollink, L., Nguyen, G., Schreiber, G., Wielemaker, J., Wielinga, B., Worring, M. Adding spatial semantics to image annotations. Workshop of Language and Semantic Technologies to support Knowledge Management Processes. EKAW 2004. 2004

[5] Cohn, A.G., Hazarika, S.M. Qualitative Spatial Representation and Reasoning: An Overview. Fundamenta Informaticae 43, pag. 2-32. 2001

[6] Galton, A., Qualitative Spatial Change. Oxford University Press, Inc., New York. 2000

[7] Srihari, R.K., Zhang, Z. Show&Tell: A semitautomated image annotation system. IEEE Multimedia. July-September 2000, 61-71. 2000

[8] MPEG Requirements Group. "MPEG-7 Overviwe", Doc. ISO/MPEG N2727, MPEG Palma de Mallorca Meeting, October 2004.

[9] Skiadopoulos, S., Koubarakis, M., Composing Cardinal Directions Relations. Artificial Inteligence vol. 152 (143-171). 2004

[10] Hollink, L., Nguyen, G., Schreiber, G., Wielemaker, J., Wielinga, B., Worring, M. Adding spatial semantics to image annotations. Workshop of Language and Semantic Technologies to support Knowledge Management Processes. EKAW 2004. 2004

[11] Antani, S., Lee, D.J., Rodney-Long, L., Thoma, G.R., Evaluation of shape similarity measurement methods for spine X-ray images. Visual Communication & Image Representation, vol. 15 (285-302). 2004

[12] Denman, H., Rea, N., Kokaram, A. Content Based Analysis for Video from Snooker Broadcasts. Lecture Notes in Computer Science. Volume 2383. Ed. Springer. Berlin. 2002

[13] Assfalg, J., Bertini, M., Colombo, C., Del-Bimbo, A., Nunziati, W. Semantic annotation of soccer videos: automatic highlights identification. Computer Vision and Image Understanding. 2003

[14] Andrade, E.L., Woods, J.C., Khan, E., Ghanbari, M. Region-based analysis and retrieval for tracking of semantic objects and provision of augmented information in interactive sport scenes. IEEE Transactions on Multimedia. Vol. 7, Issue 6. pp1084-1096. 2005

[15] G. Van Heijst, A. T. Schreiber, & B. J. Wielinga, 'Using explicit ontologies in KBS development'. International Journal of Human-Computer Studies, 45, 183-292, 1997

[16] Gonzalez, Woods. Digital Image Processing. 2nd Edition. Ed. Prentice Hall. 2002

Ad-Hoc and Personal Ontologies:
A Prototyping Approach to Ontology Engineering

Debbie Richards

Computing Department,
Division of Information and Communication Sciences,
Macquarie University, Australia
richards@ics.mq.edu.au

Abstract. Large scale or common ontologies tend to be developed using structured and formal techniques that can be equated to the Waterfall system development life cycle. However, in domains that are not stable or well-understood a prototyping approach may be useful to allow exploration and communication of ideas. Alternatively, the ontology may be part of an intermediate step or representation that provides structure, organization, guidance and semantics for another task or representation. Given that the ontology is not the end goal and possibly not reusable, the overhead of developing or maintaining such ontologies needs to be minimal. This paper reviews some of the research using ad-hoc, one-off and, sometimes, throw away, personal ontologies and provides an example of a simple technique which uses Formal Concept Analysis to automatically generate an ontology as needed from a number of data sources including propositional rule bases, use cases, historical cases, text and web documents covering a range of applications and problem domains.

Keywords: Personal Ontology, Formal Concept Analysis, Ontology Engineering.

1 Introduction

Agile software development techniques such as Rapid Application Development (RAD) and extreme programming (XP) have become an accepted way of developing software systems where requirements are not well understood and where evolution and change are the norm for that application domain. Elements of these techniques include: short cycles and continuous integration to produce and refine one or more prototypes; a failure-driven approach involving test-first programming and design; refactoring to improve structure and the use of collective ownership which is achieved by ongoing collaboration between stakeholders and pair programming [26]. Agile software development offers a major alternative to the traditional Waterfall system development life cycle model and process. While some fear that agile methods may not scale up or may result in build-and-fix models, some organisations and/or development teams clearly prefer the flexibility and speed offered by agile methods even where the requirements are well understood up-front.

A. Hoffmann et al. (Eds.): PKAW 2006, LNAI 4303, pp. 13–24, 2006.
© Springer-Verlag Berlin Heidelberg 2006

Similarly in knowledge engineering a case can be made for using techniques which develop ad-hoc and personal ontologies, which can be likened to an evolutionary or even throwaway prototype, as an alternative or exploratory precursor to the development of large-scale and/or common ontologies. This is almost the opposite to approaches which use personal ontologies to extract, restrict or guide an individual's usage or access to a larger ontology. For example, the work of Haase et al [18] allows the user to interact in *usage* or *evolution* mode with the ACM Topic Hierarchy, a domain ontology in Bibster. Usage mode restricts the user's view of the domain ontology to the topics the user has chosen to include in their personal ontology, while evolution mode allows the ontology to be extended for the individual. As [18] points out, this raises issues of management of the changing ontology and thus their work provides various change and alignment operations.

Approaching from the other direction, Chaffee and Gauch [7] ask the user to build a personal ontology in the form of a tree containing at least ten nodes and five pages per node (the goal of the ontology is to assist with web navigation) to represent their view of the world. The personal ontology is then mapped to a reference or upper level ontology. Similarly, the SemBlog personal publishing system uses a "loose and bottom-up ontology" based on a hierarchy of categories defined by the user on the basis that "everyone has those categories" which they "routinely [use to] classify ... contents to the category" [28, p. 601].

Some approaches provide technical assistance for personal ontology development. Carmichael, Kay and Kummerfield [5] use the Verified Concept Mapping technique based on concept mapping commonly used in education and for knowledge elicitation. The system contains a number of semantic concepts. These concepts are shown to the user and may be used as building blocks to develop a personal ontology. Additionally the system allows the user to define their own concepts and add these to the model, but the system will not understand the semantics of user-defined concepts. Likewise, OntoPIM [21] uses a personal ontology to assist the user to manage their personal desktop information. The personal ontology is developed by providing a Semantic Save function which allows capture of domain independent as well as domain specific metadata when an object, such as a picture or a document, are saved. Following this step, concepts are automatically mapped into the personal ontology by the system.

Sometimes adhoc and temporary ontologies are used for translating from one representation to another. For example, Moran and Mocan [27] created an adhoc ontology equivalent to an XML schema to be used by a Web Service Description Language (WSDL) description to translate from XML and the Web Services Modeling Language (WSML).

In contrast to all of the aforementioned research, this paper looks at the use of personal and ad-hoc ontologies to enable understanding of the domain to be gained and enhanced, just as one would build a throwaway or evolutionary prototype to better or incrementally understand the system requirements, application domain or test a design solution. In knowledge engineering, repertory grids [36] and Protégé [14] have been used to aid the user to discover and develop their own knowledge in a domain. In some cases the systems built acted as a communication channel to share knowledge even though the end product may have never been deployed. Personal systems, and this includes ontologies, are often more acceptable to users as they tend to be more relevant and meaningful for the individual and allow the user to use their own terminology and structure according to the users context and preferences. However, unlike the use of Protégé or repertory grids, the ontology development approach described in the next section automatically generates ontologies from other sources. When changes to the sources occur, the ontologies are simply regenerated. Such a strategy is acceptable if maintenance and ongoing reuse of the ontology is not required, as in the case of a throwaway or exploratory prototype or model.

In various projects over the past decade, Formal Concept Analysis (FCA) [40] has been used to build domain specific, personal and/or shared, ad-hoc and usually throw away ontologies from a number of alternative sources including propositional rules, cases, use cases, software specifications, web documents and keywords. FCA achieves this through the notion of a concept as a basic unit of thought comprising a set of objects and the set of attributes they share, thus providing an intensional and extensional definition for each primitive concept. FCA then applies various algorithms based on lattice and set theory to generate new concepts and allow visualization of the consequences of partial order. Section 2 of this paper provides an example of the technique. Section 3 introduces some of the applications. Conclusions are given in section 4.

2 Automatic Generation of Ontologies

FCA generally relies on the definition of exemplars or stereotypical cases to find concepts. A crosstable where each object is a row and each attribute is a column can be created to provide a formal context for the cases. The notion of a formal context reflects the view that knowledge only holds within the context it is defined. An object which has a particular attribute is marked with an "X" in the corresponding cell, see the representation in Table 1. Using FCA we are able to perform a closure operation on each object to automatically find all formal concepts for a given formal context.

What constitutes an object or an attribute depends on the data to be explored. An object could be a sentence, with the attributes comprising of the (key)words and word phrases found in a use case description or a web-based document. In the following example, a rule base has been used as input. A benefit of starting with rules is that the attribute space has been reduced to the salient features in the cases. With this input type, the rules are the objects where the rule conditions are the attributes and the rule conclusion provides the classification or label for the object. As shown in Table 1, the rules relating to the Cendrowska's contact lens dataset [6] have been used as input to the formal context. Treating each rule condition, which is really an attribute-value pair, as an attribute is similar to the technique known as *conceptual scaling* [17] which has been used to interpret a many-valued context into a (binary) formal context. The crosstable shown was automatically generated from the rules and thus did not require an additional translation step or human effort. Note that any propositional KB can be converted to a decision table [12] and therefore used in the ontology generation approach presented.

Table 1. Formal Context of Contact Lens Rules given in [6]

	1=1[1]	astigmatic = no	Tear_prod = normal	Age = Presbyopic	Prescription = myope	astigmatic = yes	age = young
Rule 0-Lens=None	X						
Rule 1 Lens=Soft	X	X	X				
Rule 2 Lens=Hard	X		X		X	X	
Rule 3 Lens=Hard	X		X			X	X
Rule 4 Lens=None	X	X	X	X	X		

The set of concepts are derived from the formal context in Table 1 by treating each row as an (object) concept and generating additional higher level concepts by finding the intersection of sets of attributes and the set of objects that share the set of attributes. For example, rules 1-4 (last four rows in Table 1) share the attribute: tear_prod=normal. This forms a new concept as shown in concept 2 in Fig. 1. Once all concepts have been found, predecessors and successors are determined using the subsumption relation ≥. This allows the complete lattice to be drawn. Disjunctions of conditions and negation must be removed to allow the rules to be converted into a binary crosstable. Fig. 1 shows the concept lattice for the Contact Lens Prescription domain. To find all attributes (rule conditions) and objects (rule numbers and conclusion codes in our technique) belonging to a concept, traverse all ascending and descending paths, respectively. For example, concept 7 in Fig. 1 includes the rule conditions (attributes) {prescription=myope, tear_production=normal, 1=1} and objects {4-%LENSN (i.e. rule 4, Lens=None) rule, 2-%LENSH (i.e rule 2, Lens=Hard)}.

From this example we can start to explore the relationships between the rules to improve our understanding of the domain. Table 1 has been provided for explanation of the approach, however, the user of this ontology development technique would not be required to define or work with the crosstable. From the user's point of view, they would firstly select the knowledge base or dataset of interest and then select which parts of the knowledge base or dataset that they wished to explore. This could be achieved via specifying one or more key words that are used to automatically select all cases or rules which contain the keywords. As the Cendrowska knowledge base is very small, all rules have been included. By looking at Fig. 1 we see the importance of the tear_production=normal concept and deduce that if we see a case where tear_production is not =normal then the default recommendation of "no lens" will be given. Alternatively, the absence of a condition covering the abnormal state may prompt the user to consider whether the default rule is adequate or whether an alternative or additional recommendation should also be given such as treatment=tear_duct_operation. Moving further down the lattice we can see that astigmatic is an important feature that will affect the prescription. If astigmatic=no then a soft lens is recommended, but only

---

[1] 1=1 ie the default condition that is always true. Rule 0 is the default rule, which will be true if no other rules fire. That is, prescription =no Lens unless ......

when the age=presbyopic and prescription=myope conditions are not true (concept 4 shows the exception rule stated in rule 4). If astigmatic=yes and the prescription=myope or age=young then a hard lens is recommended. While it is true that there is nothing shown in the concept lattice that can not be extracted from manually analysing the rules it should be apparent that the relationships between the rule conditions and conclusions are more structured and easily determined in the lattice. As in this example, the increased clarity can be useful in identifying knowledge that is potentially missing. The ontology has served as a means of understanding the domain and perhaps for validating/updating the Cendrowska knowledge base which was presumably used to provide expert opinion.

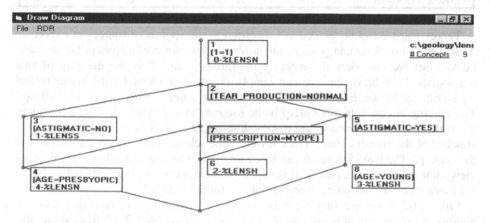

Fig. 1. The Diagram Screen in our tool which shows the Concept Lattice for the Formal Context in Table 1

The Cendrowska contact lens dataset and rules have been used by a number of researchers to demonstrate the value of various knowledge representations such as PRISM [6], INDUCT [15] and the Visual Language supported by Personal Construct Psychology [16]. We note that this data and knowledge is now out of date since the introduction of Rigid Gas Permeable (RGP) contact lenses have made hard contact lenses almost obsolete. Improvements have also been made to soft lenses in the past decade. Based on a recently expressed viewpoint of an expert optician[2] we demonstrate how new knowledge can be added to a crosstable for comparison with the old knowledge. The purpose of this comparison may be to determine if any conflicts have arisen and whether the original knowledge base needs updating. We do not include every attribute that could have been used. The rules created (i.e. the rows) and shown in Table 2 are our interpretation of the information given in the article and are not based on the use of a machine or human built set of rules developed from cases. We assume that had the optician's client data been available there would be multiple rules with exceptions to cover the four possible classifications. However, our technique is adequate for the purposes of demonstrating how knowledge from multiple sources of expertise can be displayed and reconciled using an FCA built ontology.

---

[2] http://www.epinions.com/well-review-5196-AFCVA2d-394701a9-prod2

Table 2. Formal Context of Contact Lens Rules identified from[1]

	Uncomf ortable	Visual acuity=high	Price=low	Custom Made=yes	Age = Presbyopic	Healthy= Yes	astigmatic = yes	1=1[1]
V2-Rule 0-Lens=None								X
V2-Rule 1 Lens=Soft			X		X	X	X	X
V2-Rule 2 Lens=Hard	X	X			X		X	X
V2-Rule 3 Lens=RGP	X	X		X	X	X	X	X

Tables 1 and 2 can be viewed as two different viewpoints or sources of expertise. As shown in Fig. 1 we can generate an ontology in the form of a concept lattice using FCA. Thus we can view the merging of Tables 1 and 2 as the merging of two ontologies. To achieve this we can simply combine the two formal contexts and regenerate the lattice. Fig. 2 has been developed in a newer tool, known as ConExp[3]. The reading of concepts in ConExp is the same as in our system in Fig. 1 however, the nodes are colour coded where a non-white upper semicircle indicates an attribute attached to the concept, and a black lower semicircle indicates an object attached to the concept. The two viewpoints can be distinguished by the object label: the original viewpoint does not indicate which viewpoint it belongs to, the concepts belonging to the more recent viewpoint have objects labels starting with V2.

From Fig. 2 we see that there is now an inconsistency in that soft lens are appropriate if the patient is astigmatic (Viewpoint 2/Rule 1:V2-R1:SoftLens) but also that soft lens are appropriate if the person is not astigmatic (R1:SoftLens). Either the condition is no longer important for prescribing or choosing which lens is appropriate or the rules shown are incorrect. Perhaps R1 is no longer true, or perhaps V2-R1 needs to be further qualified to state that only Toric soft lenses are appropriate for astigmatic patients. As I am not an optician, I can not answer this question. There are further questions prompted by Fig. 2 that I would like to ask an optician, or to investigate myself through review of the literature, such as whether normal tear production is still a (pre)condition of prescribing any contact lens as appeared to be the situation from Cendrowska's dataset. Clearly the ontology opens up a communication channel for a novice such as myself to explore the domain further with or without a domain expert's assistance and also a means of assisting the domain expert to think about and articulate their knowledge about the domain.

The approach offered here is a human in the loop approach where the goal of the approach is to allow identification of similarities and differences between sources to be highlighted and reconciled via discussion. In a number of projects, we have extended the ontology merging process with a number of reconciliation strategies that assist in identifying if two concepts are in a state of contrast, correspondence, conflict or consensus and a means to identify the degree of consensus between two or more ontologies. See [3, 29, 30] for details. A more automated approach to bottom-up ontology merging can be achieved with the FCA-MERGE technique [36].

---

[3] http://www.sourceforge.net/projects/conexp

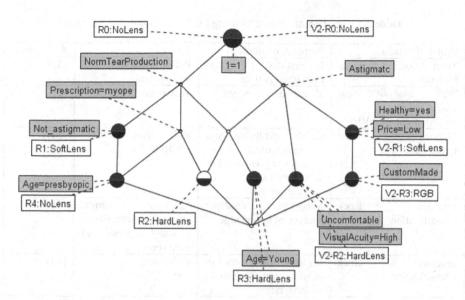

Fig. 2. The Diagram Screen in ConExp which shows the Concept Lattice for the Formal Contexts in Tables 1 and 2

3 Applications of Adhoc and Personal Ontologies

Common ontologies seek to provide a reusable library of concepts. Adequate time and effort needs to be taken to get the concepts right, define the terms, relationships, axioms and so on. The approach to development is typically top-down. FCA however allows us to work from bottom up, with minimal modeling of the domain beyond the creation of a crosstable containing objects and attributes. As demonstrated in the previous section, what is modeled as an object or an attribute can vary according to the input or questions to be explored. Table 3 provides a summary of a number of projects that have used FCA derived ontologies using input other than conventional cases for a range of different purposes. In each of the projects, the interplay of FCA and ontologies has provided a learning technique, allowed analysis and navigation of the derived ontology and the ontology has enhanced the FCA application [10]. The list of projects is far from exhaustive but gives a taste of the possibilities.

Elsewhere discussion can be found on the nature of the ontology developed using FCA (e.g. [10, 19]) together with a comparison of other techniques for ontology development (e.g. [32]). The purpose of this paper is to consider the role and value of using a domain and/or individual specific ontology as a communication channel, or alternatively, a mediating or temporary representation.

Whether FCA is used to compare rules in a knowledge base, use case descriptions or documents, the approach allows and encourages individuals to express themselves using their own terms and on their own without the interference and restrictions associated with group thinking. This has the benefit of increased engagement with the

Table 3. Generation of ontologies using FCA from non-standard cases

Project/Domain	Purpose of Ontology	Input	Reference
Treat Anorexia Nervosa Patients	Understand the individual patient	Personal construct and survey data	[37]
Web analysis and visualisation environment (WAVE)	Visualise Web pages	Web pages	[2]
Management of the Lotus Crop	1. The evolution and management of emerging knowledge 2. Combination of multiple KB	Rules from multiple MCRDR KB	[34]
Igneous Rock Classification	Reconcile multiple sources of knowledge	Laddered grids, structured interviews, protocol analysis and card sorts	[31]
Igneous Rock Classification	Build an initial model to provide initial domain understanding	Card sorts	[13]
The retrieval of web-based documents	Develop personal ontology to make retrieval more customized to the user	User defined categories	[24]
Requirements Engineering	1. Reconciliation of stakeholder viewpoints. 2. More complete use case description	Use case descriptions	[3, 32]
Software Engineering	Validation of models	Software Specifications	[39]
The retrieval of web-based documents	Structure key words, structure relevant documents	Key words, web documents	[8, 9, 25]
Pathology	Discovery of higher level knowledge and patterns	Knowledge bases	[35]
Travel Text Corpus	Merging Ontologies	Web Documents	[38]
Tacit Knowledge Measurement	Compare Responses to Scenarios in a Tacit Knowledge Inventory	Survey Data	[4]
Scientific Knowledge Management	Analyse the value of FCA in knowledge technologies and ontology building	Lists of PC Members, session topics, publications, entity relational data model	[19]

task and ownership of the knowledge. This becomes even more important when the goal is to build a shared model. By starting with separate sources each group member owns and defends a viewpoint to provide a truly representative and more complete final model. Just as a prototype developed with a 4GL may not give the developer as much freedom and control over the application developed at they might like, the end user can see results sooner and may even be able to use the 4GL to develop the system themselves. As in software engineering, knowledge and ontology engineering may benefit from simpler and user-definable languages, but perhaps with the tradeoff of expressivity. However, creating more expressive ontology languages does not necessarily result in humans being better able to express themselves as the learning curve and structure is usually greater and more complex.

Another plus is the use of the concept lattice for diagrammatic reasoning. For example, our evaluation studies showed that participants were able to identify similarities and differences between viewpoints more quickly and accurately than when presented with the same information in its original textual format [29, 31].

The FCA notion of a concept is compatible, though differences usually exist, with other concept processing approaches. Many of the projects combined FCA with other techniques such as language technology [3, 11, 30, 31], information retrieval [8, 9], Description Logics [29], Conceptual Graphs [41] and Knowledge Based Systems [33]. In the study by Spangenberg and Wolf [37] FCA was combined with the use of Personal Construct Psychology [22] and survey techniques. That study used the repertory grid approach to elicit the responses of anorexia nervosa patients to various people in their lives based on a number of bipolar scales. The goal of that work was to assist the physician to identify the issues faced by the individual patient and did not seek to determine a shared view across patients, which would have been irrelevant in this context, unless of course the patients were related to one another. This early work demonstrated the value of the FCA lattice to provide a personal ontology. To require these patients or the doctor to follow more conventional ontology building processes, typically requiring the assistance of a knowledge engineer, would be out of the question. The physician would probably be unwilling to undergo the special training needed and the presence of a knowledge engineer would interfere with the elicitation process and breach patient-doctor confidentiality.

4 Discussion and Conclusion

We note that there is some conflict between the goal of ontologies and their actual usage. Some have argued that Gruber's definition has led to a view that ontology is :

> " 'a model' where what is being modeled are the concepts or ideas people have in their minds. This reductive error has its roots in the recent tendency to use the word 'ontology' to mean little more than a controlled vocabulary with hierarchical organization"[4].

When one remembers that ontologies are concerned with the nature of being and the world, the focus on what is in people's heads or the words they use fits more with linguistics, psychology or epistemology rather than metaphysics.

We also note that the goal to create large-scale common ontologies sought to address the KA bottleneck by providing guidance and allowing sharing and reuse. However, it is unclear whether ontology engineering has simply moved the bottleneck higher and earlier in the KA process. Also the desire to share and reuse has led to the need for strategies for merging and reconciliation.

---

[4] http://ontologyworks.com/what_is_ontology.php

Currently, ontologies are seen to play a pivotal role in the Semantic Web, together with semantic markup languages. However, the effort involved in the two-step authoring and annotation process in a formalism such as OIL and/or RDF "tends to reintroduce the impulse to set up the 'right' ontologies in advance. This seems contrary to letting 'anyone say anything' [2] or, perhaps, it simply raises the burden of generating Semantic Web content to an inhibitory level" [20]. To address this issue, Kalyanpur et al [20] offer the Semantic Markup Ontology and RDF Editor (SMORE) to support adhoc ontology use, modification, combination and extension. However, what SMORE attempts to achieve is a more seamless environment which merges and simplifies authoring and annotation. The approach allows adhoc use of ontologies, not to be confused with the use of adhoc ontologies as addressed in this paper. However, similar to SMORE, we seek to offer a practical approach.

Bennett and Theodoulidis [1] have investigated the notion of personal ontology and its relationship to organizational ontology and knowledge. They see that a personal ontology is the outcome of personal world experiences leading to personal knowledge that forms a personal ontology. When individuals begin to share their ontologies and agree on meanings, organizational ontologies begin to emerge. In contrast to the flow from experience to knowledge to ontology for the individual, at the organizational level, once an organizational ontology exists, organizational knowledge can emerge resulting in experiences at the organizational level which feed back into personal experiences. If such a cycle does exist, it may be necessary to ensure that approaches for engineering common, upper or reference ontologies support the ongoing development, sharing and integration of personal ontologies.

Despite its various shortcomings, the Waterfall system development life cycle is still the main development method used in many organizations. In practice the method is often modified to include incremental and iterative cycles within and between certain phases. Likewise, common ontologies, such as CYC or WordNet are in widespread use and are often used in conjunction with domain specific and sometimes personal ontologies. The use of FCA to rapidly develop domain and personal ontologies offers many parallels to agile software development in that the technique is incremental, rapid, collaborative and the development cycle is essentially test-first driven producing prototypes (adhoc ontologies) for exploring domain or individual-specific concepts. Using FCA to automatically generate an ontology from whatever source can be mapped to a crosstable with minimal effort, becomes attractive particularly where the ontology or domain itself is volatile and/or temporary.

References

[1] Bennett, B. R. and Theodoulidis, B., Towards a notion of Personal Ontology, http://citeseer.comp.nus.edu.sg/44041.html, accessed 10th July 2006.

[2] Berners-Lee, T. Handler, J. and Lassila. O. (2001) *The Semantic Web*. Scientific American, May 2001.

[3] Boettger, K. Schwitter, R., Mollá, D. and Richards, D. (2003) Towards Reconciling Use Cases via Controlled Language and Graphical Models In O. Bartenstein, U. Geske, M. Hannebauer, O. Yoshie (eds.), *Web-Knowledge Management and Decision Support*, LNCS, Vol. 2543, pp. 115-128, Springer Verlag, Heidelberg, Germany.

[4] Busch, P. and Richards, D. (2004) Modelling Tacit Knowledge via Questionnaire Data, *Proc.of 2nd Int.Conf.on FCA (ICFCA 04)*, Feb 23-26, 2004, Sydney, Australia, 321-329.

[5] Carmichael, D J, J Kay, R J Kummerfeld, (2004) Personal Ontologies for feature selection in Intelligent Environment visualisations, in Baus, J, C Kray and R Porzel, *AIMS04 - Artificial Intelligence in Mobile System*, 44-51.

[6] Cendrowska, J. (1987) An algorithm for inducing modular rule. *Int. Journal of Man-Machine Studies* 27(4):349-370.

[7] Chaffee, J. and Gauch, S. (2000) Personal Ontologies For Web Navigation. In Int.Conf. Info. Knowledge Mgt (CIKM), pp. 227-234.

[8] Cho, W. C. and Richards, D. (2004) Improvement of Precision and Recall for Information Retrieval in a Narrow Domain: Reuse of Concepts by Formal Concept Analysis, *Proc. IEEE/WIC/ACM Int. Conf. Web Intell. (WI'2004)*, Sept. 20-24, Beijing, China, 370-376.

[9] Cho, W. C. and Richards, D. (2006) Automatic construction of a concept hierarchy to assist Web document classification, *Proc. 2nd Int.Conf.on Info. Mgt and Business (IMB.2006)*, 13-16 February, 2006, Sydney, Australia.

[10] Cimiano, P., Hotho, A., Stumme, G. and Tane, J. (2004) Conceptual Knowledge Processing with Formal Concept Analysis and Ontologies, LNCS, Vol 2961:189 – 207.

[11] Cimiano, P., Staab, S. and Tane, J. (2003) Automatic Acquisition of Taxonomies from TexT: FCA meets NLP, Proc. of the Int. W'shop on Adaptive Text Extraction and Mining.

[12] Colomb, R.M. (1989) Representation of Propositional Expert Systems as Decision Tables *Technical Report TR-FB-89-05* Paper presented at 3rd *Joint Aust. AI Conf.* (AI'89) Melbourne, Victoria, Australia 15-17 November, 1989.

[13] Erdmann, M. (1998) Formal concept analysis to learn from the sisyphus-III material.In: Proc. of 11th KA for KBS Workshop, KAW'98, Banff, Canada 1998.

[14] Eriksson, H., Fergerson, R. W., Shahar, Y. and Musen, M. A. (1999) Automatic Generation of Ontology Editors In *KAW'99*, 16-21 October, 1999, Banff.

[15] Gaines, B.R., (1991) Induction and Visualization of Rules with Exceptions In J. Boose & B. Gaines (eds), *Proc.6th Banff AAAI KAW'91*, Canada,Vol 1: 7.1-7.17.

[16] Gaines, B. R. and Shaw, M.L.G. (1993) Knowledge Acquisition Tools Based on Personal Construct Psychology *Knowledge Engineering Review* 8(1):49-85.

[17] Ganter, B. and Wille, R., (1999) Formal Concept Analysis – Mathematical Foundations, Springer-Verlag, Berlin.

[18] Haase, P. Stojanovic, N., V˚olker, J. and Sure, Y. (2005) Personalized Information Retrieval in Bibster, a Semantics-Based Bibliographic Peer-to-Peer System, *Proceedings of I-KNOW '05 Graz, Austria, June 29 - July 1, 2005*

[19] Kalfoglou, Y, Dasmahapatra, S. and Chen-Burger, Y-H, (2004) FCA in Knowledge Technologies: Experiences and Opportunities, *LNCS* 2961, Feb 2004, Pages 252 – 260

[20] Kalyanpur, A., Parsia, B., Hendler, J. and Golbeck, J. (2001) "*SMORE - Semantic Markup, Ontology and RDF Editor*". Technical Report www.mindswap.org/~aditkal/ SMORE.pdf

[21] Katifori, V., Poggi A., Scannapieco, M., Catarci, T. and Ioannidis, Y. (2005) OntoPIM: How to Rely on a Personal Ontology for Personal Information Management, In *Proc. of the 1st Workshop on The Semantic Desktop*, 2005.

[22] Kelly, G.A, (1955) *The Psychology of Personal Constructs* Norton, New York.

[23] Kent, R.E. and C. Neuss. 1995. Creating a Web Analysis and Visualization Environment. Computer Networks and ISDN Systems, 28.

[24] Kim, S., Hall, W., and Keane, A. (2001). Using document structure for personal ontologies and user modeling. In Bauer, M., Vassileva, J., and Gmytrasiewicz, P. (Eds.), *User Modeling: Proc. of the 8th In. Conf. UM2001*, Berlin: Springer, pages 240–242.

[25] Kim, M. and Compton, P., (2004) Evolutionary Document Management and Retrieval for Specialized Domains on the Web. Int.l Jrnl of Human Computer Studies. 60(2):201-241.

[26] Maciaszek L.A., Liong B.L., (2005), *Practical Software Engineering - A Case Study Approach*, Addison-Wesley.

[27] Moran M. and Mocan A. (2005) Towards Translating between XML and WSML based on mappings between XML Schema and an equivalent WSMO Ontology: *Second WSMO Implementation Workshop (WIW '05)*, June 2005, Innsbruck, Austria.

[28] Ohmukai, I, Takeda, H., Hamasaki, M., Numa, K. and Adachi, S. (2004) Metadata-driven Personal Knowledge Publishing. S.A. McIlraith et al. (Eds.): *ISWC 2004*, LNCS 3298, Springer-Verlag Berlin Heidelberg, 591–604.

[29] Prediger, S. and Stumme, G. (1999) Theory-driven Logical Scaling: Conceptual Information Systems meet Description Logics. *KRDB 1999*: 46-49

[30] Richards, D. (1998) An Evaluation of the Formal Concept Analysis Line Diagram, *Poster Proc. AI'98*, 13-17 July 1998, Griffith University, Brisbane, Australia, 109-120.

[31] Richards, D. (2000) Reconciling Conflicting Sources of Expertise: A Framework and an Illustration, Proc. of PKAW'2000, December 11-14, 2000, Sydney.

[32] Richards, D. (2003) Merging Individual Conceptual Models of Requirements, *Special Issue on Model-Based Requirements Engineering for the Int. Jrnl of Requirements Engineering*, (2003) 8:195-205.

[33] Richards, D. (2004), Addressing the Ontology Acquisition Bottleneck through Reverse Ontological Engineering, *Jnl of Knowledge and Information Systems* (KAIS), 6:402-427.

[34] Richards, D. and Compton, P. (1997) Uncovering the Conceptual Models in Ripple Down Rules In Dickson Lukose, Harry Delugach, Marry Keeler, Leroy Searle, and John F. Sowa, (eds) (1997), *Conceptual Structures: Fulfilling Peirce's Dream, Proc. of the 5ᵗʰ Int. Conf. on Conceptual Structures (ICCS'97)*, LNCS 1257, Springer Verlag, Berlin, 198-212.

[35] Richards, D. and Malik, U. (2003) Multi-Level Knowledge Discovery in Rule Bases, *Applied Artificial Intelligence*, Taylor and Francis Ltd, March 2003, 17(3):181-205

[36] Shaw, M.L.G. and Gaines, B.R., (1991) Using Knowledge Acquisition Tools to Support Creative Processes In *Proc.of the 6th KA for KBS Workshop,* Banff, Canada.

[37] Spangenberg, N., Wolff, K.E. (1988) Conceptual grid evaluation, In *Classification and related methods of data analysis*. Spangenberg, Norbert and Karl Erich Wolff, Eds., Amsterdam, North-Holland: 577-580.

[38] Stumme, G. and Madche., A.(2001) FCA-Merge: Bottom-up merging of ontologies. In 7th Intl. Conf. on Artificial Intelligence (IJCAI '01), pages 225--230, Seattle, WA, 2001.

[39] Tilley.T. (2003) A Software Modelling Exercise using FCA. In *Proceedings of the 11ᵗʰ Inter-national Conference on Conceptual Structures (ICCS'03), Springer LNAI 2746, Dresden, Germany*, July 2003.

[40] Wille, R. (1992) Concept Lattices and Conceptual Knowledge Systems *Computers Math. Applic.* (23) 6-9: 493-515.

[41] Wille, R. (1997) Conceptual Graphs and Formal Concept Analysis. In D. Lukose et.al., editor, *Conceptual Structures: Fulfilling Peirce's Dream*, Springer, LNAI 1257, 290--303.

Relating Business Process Models to Goal-Oriented Requirements Models in KAOS

George Koliadis and Aditya Ghose

School of Information Technology and Computer Science,
University of Wollongong,
Wollongong, N.S.W. Australia
{gk56, aditya}@uow.edu.au
http://www.uow.edu.au

Abstract. Business Process Management (BPM) has many anticipated benefits including accelerated process improvement, at the operational level, with the use of highly configurable and adaptive "process aware" information systems [1] [2]. The facility for improved *agility* fosters the need for continual *measurement* and *control* of business processes to assess and manage their effective evolution, in-line with organizational objectives. This paper proposes the GoalBPM methodology for relating business process models (modeled using BPMN) to high-level stakeholder goals (modeled using KAOS). We propose informal (manual) techniques (with likely future formalism) for establishing and verifying this relationship, even in dynamic environments where essential alterations to organizational goals and/or process constantly emerge.

1   Introduction

Business Process Management (BPM) in its "third wave" [1] has been conveyed as: enabling intelligent business management [3]; facilitating the redesign and organic growth of information systems [4]; and, obliterating the business - IT divide [1]. Business processes undergo an evolutionary life-cycle of change. This change is brought on by the need to satisfy the constantly changing goals of varied stakeholders and adapt to the accelerating nature of change in today's business environment [5]. The need for change is best described in [6] as the transition from an initial "unsatisfactory" (i.e. as-is) state to a new hypothetically "desired" (i.e. to-be) state. The desired state is theoretically based on the assumption that it more effectively satisfies related operational goals [4] [7] [8] in-line with higher-level strategic goals. It is therefore important that the criterion for effective process change - i.e. stakeholder goals, be explicitly stated, communicated and traceable to any changes that are proposed, approved, and/or implemented.

The new-found *agility* provided by BPM, however presents the need for methods to successfully *control* and *trace* the evolution of processes. This need is affirmed in [7], by stating that organizations evolve from their original intentions through complex and unpredictable growth. BPM aims to support the evolution

A. Hoffmann et al. (Eds.): PKAW 2006, LNAI 4303, pp. 25–39, 2006.

of organizations and their processes, however controls are still needed to ensure that operational as well as higher-level goals (i.e. of more strategic concern) are continually satisfied, allowing for "organizational growth in the right direction". In order to meet goals however, there is a need to support traceability between processes and organizational goals - "You can't manage what you can't trace" [9].

We have proposed a method (GoalBPM) to support the controlled evolution of business processes. Control is supported through the explicit modeling of stakeholder goals, their relationships (be it either refinement, conflict or obstruction), and their evolution traceable to related business processes. GoalBPM is used to couple an existing and well-developed, informal-formal goal modeling and reasoning methodology - i.e. KAOS [10], and a newly developed business process modeling notation - i.e. BPMN [11]. This is achieved through the identification of a *satisfaction* relationship between the concepts represented. GoalBPM itself can be seen as an "adapter" that integrates the two models, to support their co-evolution and synergistic use.

This paper firstly presents a background to the associated domains of business process modeling and goal-oriented requirements engineering. An informal overview of the GoalBPM method is subsequently outlined with a simple example for illustration.

2   Business Process Modeling with BPMN

We have initially chosen the Business Process Modeling Notation (BPMN), developed by the Business Process Management Initiative (BPMI.org) for use in the construction of GoalBPM.

A *Business Process* is a set of dynamically co-ordinated activities controlled by a number of dependent, social participants. Processes are represented in BPMN using **flow objects**: *events* (circles), *activities* (rounded boxes), and *decisions* (diamonds); **connecting objects**: *control flow links* (unbroken directed lines), and *message flow links* (broken directed lines); and **swim lanes**: *pools* (high level boxes containing a single process), and *lanes* within pools (subboxes).

We refer to Figure 2, a public Package Sorting process, as an example to illustrate BPMN. The process requires the interaction of two high-level process participants - the Transport Organization, and a Transport Authority. Collaboration between participants on the model is represented by *message flow links* between *activities* within *pools*. Responsibility within the Transport Organisation is delegated to two roles - the Sort Operations, and a Bond Operations. Responsibility assignment within a pool is represented using *lanes* (i.e. pool divisions). Each pool within a process model represents a single process. Processes are initiated by a *start event*, represented as a circle at the beginning of each pool. *Control flow links* between activities, decisions and events, represent the controlled progression through each process. A *decision gateway*, (i.e. diamond) can be seen in the figure, identifying the need to make a choice on whether to bond a package. Finally, the process is completed with an *end event*, or bold circle toward the end of a process.

3   Goal-Oriented Requirements Engineering with KAOS

We have chosen a GORE method that is focused toward both the early and late phases of RE, specifically KAOS (Knowledge Acquisition in autOmated Specification of software systems) [10], to represent the organizational goals related to business process execution.

Goal Declaration in KAOS. Goals are declared in terms of desired, timely effects within a composite system (e.g. *Achieve[MeetingScheduled]*). Goals are conceptually modeled in KAOS on a semantic net that represents a hierarchy of parent goals and their refinements into sub-goals. Goals that exist higher in the hierarchy represent the high-level goals (i.e. *strategic* concerns) of the organization. These goals are not "clear cut", in the sense that their satisfaction is complex and cannot be proven without common interpretation. These goals are then *refined* down the hierarchy into sub-goals that are more *operational* in nature. That is, their assignment to a small group of individuals responsible for a number of operational activities illustrates the means by which they are satisfied.

Goals can be either 'AND' or 'OR' refined (Figure 3). An 'AND' refinement of a goal states that the parent goal is satisfied if all the goals in the refinement are satisfied. An 'OR' refinement on the other hand states that the parent goal is satisfied if a single refinement is satisfied. This allows for the modeling of alternative refinements for goal satisfaction. KAOS also provides a criterion for halting the refinement process, in that if a goal can be assigned the sole responsibility of a single environmental role (i.e. agent in the composite system); there is no need for further goal refinement to occur. This also provides a means by which to make the transition between goals and the constrained operations that satisfy those goals [12].

Goal Definition in KAOS. KAOS supplies an optional formal assertion layer that allows for the specification of goals in Real-Time Linear Temporal Logic (RT-LTL) [10] [13]. These formal goal assertions allow for precise specification of goals, as well as supporting the use of developed formal reasoning techniques that aid in identifying/resolving conflicts between goals and proving absolute/partial goal satisfaction.

A formal goal definition in KAOS begins with the assertion of the objects the goal *concerns*. In KAOS, these objects are declared in the object model. The definition then states the desired temporal ordering of states the concerned objects must hold in order to *satisfy* the goal.

Goals are defined in the form:

$$C \Rightarrow op\mathrm{T},$$

KAOS Goal Modelling

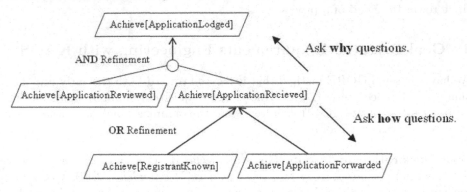

Fig. 1. Modeling goals in KAOS

where C and T are assertions about environmental situations (i.e. current and target), and *op* is a temporal operator that signifies the desired temporal nature of the target situation (T), in relation to a current situation (C). This paper requires knowledge only of the two most used operators. Namely, 'at *some* time in the future' denoted by a open diamond '◇', and 'at *all* times in the future' denoted by an open square '□'. For a complete list of the operators used in KAOS, see [10].

Take for example, the 'PackageSortedToDestination' goal below:

Goal Achieve[PackageSortedToDestination]
InformalDef If a package is received at a sort facility, then the package will eventually be forwarded to its known destination.
FormalDef ∀p: Package, sf: SortFacility
Received(p, sf) ⇒ ◇Forwarded(p, p.Destination)

Also note: package destinations (i.e. p.Destination) are *attributes* of a package.

Patterns for Declaring and Defining Goals. KAOS defines a number of commonly used "Goal Patterns" that generalize the timeliness of target situations. They provide an informal method to initially declare goals, as well as to guide formal definition. *Achieve Goals* $(C \Rightarrow \Diamond T)$ desire achievement 'some time in the future'. That is, the target must eventually occur. (e.g. Achieve[PaymentRecieved]). *Cease Goals* $(C \Rightarrow \Diamond - T)$ disallow achievement 'some time in the future'. That is, there must be a state in the future where the target does not occur (e.g. Cease[Operation]). *Maintain Goals* $(C \Rightarrow \Box T)$ must hold 'at all times in the future' (e.g. Maintain[EmployeeSafety]). *Avoid Goals* $(C \Rightarrow \Box - T)$ must not hold 'at all times in the future' (e.g. Avoid[LateEntry]).

4   Linking Goal and Process Models

The GoalBPM methodology relies on establishing relationships between goal models and process models. This relationship is established in two stages. First, *traceability links* are established between goal nodes in the goal model to activities (or end events of complete processes) in the process model. Second, *satisfaction links* are established between goals and processes.

A traceability link is an informal statement of a relationship between a goal and a process (or sub-process). In establishing such a link, we are effectively asserting that the goal in question has some bearing on the process (or sub-process) under consideration. A traceability link does not necessarily lead to a satisfaction link. Sometimes, a process or sub-process may be related to a goal because it *obstructs* it, in the sense of [14]. We draw a traceability link between a goal and a process end event if the entire process in question has some bearing on the goal. We draw a traceability link between a goal and an activity if the sub-process ending in that activity relates to the goal. In general, traceability links need to be established by analysts. Some guidance can be offered in this process by using cues present in the goal and process models. For instance, the names of goals and processes (or activities) can suggest traceability links. A transport organization's operational goal to achieve "PackageSortedToDestination" can be traceable to the "Package Sorting" process within the organization.

Traceability between goals and processes can be identified through cross examination of the links between the pre/post conditions for specific processes and the pre/post conditions for specific goals. A process is made available for execution when a specific pre condition has been met (e.g. a *customer that has submit a registration form* in the 'Register New Customer' process), and completed upon meeting a post condition (e.g. the customer is validated and their details are stored). These pre and post conditions can be related to the pre and post conditions for specific organizational goals. Take for example the goal, '*all new customer registrations* require credit reporting and verification', is related to the 'Register New Customer' process by way of the pre condition.

A *satisfaction link* is a traceability link where the process (or sub-process) in question satisfies the goal involved in the link. A satisfaction link can be of two types:

- *Normative satisfaction links:* These indicate that a process or sub-process *must* satisfy the relevant goal. Such links articulate desired states of affairs.
- *Descriptive satisfaction links:* These indicate the "as-is". We obtain descriptive satisfaction links from effect annotations of processes, using techniques that we discuss in the next section.

5   Using Model Annotations to Verify Goal Satisfaction

GoalBPM establishes satisfaction links between goals and process models in three steps. First, it annotates process models with *effect annotations*. Second, it identifies a set of *critical trajectories* from a process model. Third, it identifies the subset of the set of traceability links that represent satisfaction links by analyzing critical trajectories relative to process effect annotations. The satisfaction links thus obtained are *descriptive* satisfaction links. A final step in GoalBPM is to use a comparison of the set of normative satisfaction links with the set of descriptive satisfaction links to drive the processes of *goal model update* and/or *process model update*.

Our approach may be viewed as an instance of the *state-oriented* view [15] [16] [17] of business processes as opposed to the *agent-oriented* or *workflow* views. However, we are not explicitly state-based in that we do not seek to obtain state machine models from process models, for two reasons. First, BPMN models in general do not guarantee finite state systems, making the application model checking techniques difficult. Second, the derivation of state models from BPMN models appears difficult at this time, due to the high-level, abstract nature of BPMN models.

5.1   Effect Annotations

A *process activity* (i.e. as represented in BPMN as a rounded box) is an element on a process model that indicates required state transitions in order for the process to progress toward the achievement of all related goals. The labeling of an activity (e.g. 'Register New Customer') generalizes a number of possibly desired/undesired results, or outcomes. Due to this generalization, most process models do not satisfactorily depict the lowest level state achievements required for process progression. They are too 'high-level' and do not provide an in depth understanding of the process and its ability to achieve desired goals or objectives. This understanding is important when trying to prove goal achievement, which is reliant on the achievement of certain target states. In order to provide greater understanding of process models (i.e. state transitions in particular) to support their analysis in relation to the goals they hope to satisfy, we augment the process model with 'effect annotations'.

An *effect* is the result (i.e. product or outcome) of an activity being executed by some cause or agent. It indicates the achievement of a certain environmental state communicated through an event. An *effect annotation* relates a specific result or outcome to an activity on a business process model. It explicitly states a result of the activity if the conceptual model were to be hypothetically executed. A *cause* relationship exists between a process activity and an effect (i.e. the process activity causes the effect to occur). An activity can cause many effects and an effect can be caused by a number of activities.

The manner in which an activity is executed may result in alternative outcomes. For this reason, effect annotations are related to activities in an AND/OR refinement similar to goal refinement in KAOS. This alternative execution is derived from the dynamic co-ordination represented in the process model through decision gateways (i.e. diamonds in BPMN). Decisions commit to alternate paths of execution based on the current state of the process. This is achieved through the specific effects of prior activities that have been executed. The conditions for the decision on which choice of path to commit to can help to identify important effects on prior activities in the current, or in other processes. These influential activities and their required effects for the current path of execution, need to be identified and represented along with the effects of the current process to prove goal satisfaction.

We define an *effect annotation* to include:

– a *label* that generalizes the behavior of the effect in relation to its environment (e.g. 'CustomerDetailsStored'). Whereas the labeling of an activity is made in the optative mood (i.e. a desire), an effect annotation is made in the indicative mood (i.e. a fact).ling of an activity is made in the optative mood (i.e. a desire), an effect annotation is made in the indicative mood (i.e. a fact).
– a *designation* specifying whether the effect is a 'normal' (i.e. expected) outcome for the activity that in turn aims toward goal achievement, or an 'exceptional' (i.e. unaccepted) effect that deviates from goal achievement. (e.g. 'RegistrationValidated' may be a normal outcome for a customer registration activity, whereas 'RegistrationRejected' may be exceptional)
– a *informal definition* an informal definition describing the effect in relation to the result achieved in its environment (e.g. 'The details relating to the current customer have been stored within the system.'). This provides an informal explanation (i.e. meaning) of the effect in relation to the real-world environment.
– a *formal definition* (optional) defining achieved states to aid in mapping to formal goal definitions in the chosen goal definition formalism (i.e. in this case KAOS). (e.g. '$\forall$ c: Customer, ($\exists$ cr: CustomerRecord) Stored(c.Details, cr)')

At the tool level, effect annotations can be viewed on a business process model graphically, or added to meta-information relating to the process activities. They can then be analyzed along with the process and associated goals as described in the subsequent sections.

5.2 Trajectory Decomposition

The dynamic co-ordination of activities controlled by responsible agents is represented / supported on a business process model by way of decision gateways (i.e. diamonds in BPMN). This manner of 'per instance' process control allows for the existence of many *process trajectories* through the process (i.e. possibly even an infinite number when cycles are included). We classify a single trajectory as a unique and supported sequence of activity execution. Each trajectory results in a 'cumulative effect' for the given process. We use the term trajectory to signify a specific 'chosen path' through the process that results in a specific/unique outcome or 'cumulative effect'.

During the trajectory decomposition process, specific effects need to be chosen where alternative effects are available on each activity. This choice relies on the decisions influencing the path of the particular trajectory. We choose between alternative effects based on their conformance to the current trajectory.

Business process models support and represent exceptional trajectories. These trajectories do not necessarily satisfy process goals. They react to exceptional events that occur in the process by re-routing the current path of execution so that alternative steps can be taken to either resolve the exceptional situation or abort the process. We can therefore label a trajectory as having either a *normal* or *exceptional* type, based on its final achieved state. Commonly, a trajectory that cannot resolve exceptions and requires termination prior to meeting all the goals it must satisfy, is classed as exceptional. This is discussed in the following section.

The identification of all unique process trajectories can be difficult, given the complex nature of activity interleaving and iteration possible in the process models. For this purpose, it is recommended that an automated method for deriving all possible process trajectories be available.

5.3   Goal Satisfaction and Cumulative Effect Assessment

We firstly progress through each process trajectory and compare effects with traceable satisfaction goals in the goal model. Effects are compared to the desireability and temporal ordering of effects in normative goals and descriptive satisfaction links are established as we progress through the trajectory.

We then analyze and classify each trajectory as either *normal* or *exceptional*. A normal trajectory in relation to the goal model leads to the satisfaction of all normative goals. An exceptional trajectory, as described in the previous section, satisfies a limited number of normative satisfaction goals.

In order for a satisfaction relationship to exist between a goal model and an associated process model, there must be at least one normal trajectory. That is, the process model must support at least one valid means by which to satisfy the required normative goals of the process.

Finally, we analyze the outcome of the satisfaction process, identifying whether the process supports the achievement normative satisfaction goals and classify the satisfaction relationship between the process and the associated goal model as either *strong*, *weak*, or *unsatisfied*. A *strong* satisfaction relationship is determined if all possible trajectories are 'normal' (i.e. satisfy all associated goals). On the other hand, a *weak* satisfaction relationship is said to exist when there is at least one 'exceptional' trajectory and one 'normal' trajectory. This classification, delineating between weak and strong satisfaction, can be important when evaluating the competency of the process in recovering from exceptional situations that may arise during enactment. An *unsatisfied* satisfaction relationship is the result of there not being a single 'normal' trajectory decomposed from the process. This classification requires that changes are made to either the process and/or goal model to establish a weak or strong satisfaction relationship.

6   Example

We apply the GoalBPM to a single case within a Transport Organisation for illustration. GoalBPM is specifically applied to a core operational 'Package Sorting' process within the organization (see Figure 2). Furthermore, we introduce change to the goal model, and identify inconsistencies with the current business process that need to be addressed to maintain the satisfaction relationship.

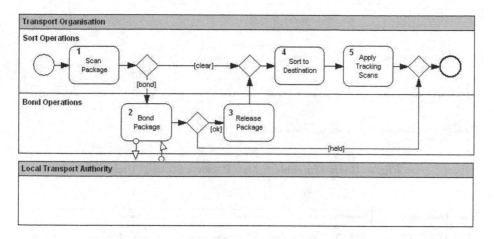

Fig. 2. The 'Package Sorting' Process of the Transport Organization

6.1   Current Business Context and Process

The operational objective the 'Package Sorting' process aims to achieve is the prompt routing of *packages*, upon arrival to a *sorting facility*, to their respective destinations. There are three primary participants in the process whose objectives must be met: *Transport Organization* - whose concern is the efficient routing of packages to their destinations by assigning responsibility to internal sort operations as well as bond operations whose role is to liaise with transport authorities for prompt delivery clearance; *Customers* - whose concern is prompt delivery and package traceability; and *Transport Authorities* - whose concern is with maintaining a high level of integrity in regards to border control through package screening. These requirements are represented on the goal model in Figure 3, with definitions supplied in Figure 4.

The 'Package Sorting' process in Figure 2, represents the current 'as-is' co-ordination of activities and interactions aimed toward achieving traceable goals. We apply GoalBPM to the process and goal model in order to evaluate the current state of the satisfaction relationship before introducing the aforementioned changes to the goal model.

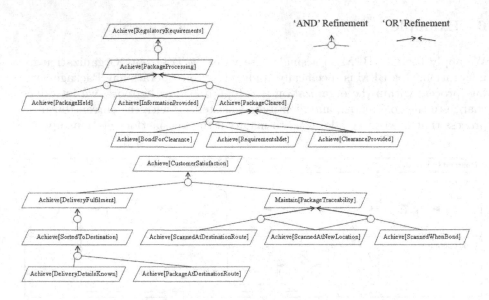

Fig. 3. Goals Traceable to the 'Package Sorting' Process

Goal Type	Goal Declaration	Concerned Objects	Antecedent	Consequent
Achieve	PackageProcessing	Package, SortFacility, TransportAuthority	Arrives(p, sf)	<>Processed(p, ta)
Achieve	InformationProvided	Package, SortFacility, TransportAuthority	Arrives(p, sf)	<>Provided(p.DeliveryDetails, ta)
Achieve	PackageCleared	Package, TransportAuthority	Provided(p.DeliveryDetails, ta)	<>Cleared(p, ta)
Achieve	PackageHeld	Package, TransportAuthority	Provided(p.DeliveryDetails, ta)	<>Held(p, ta)
Achieve	BondForClearance	Package, SortFacility, TransportAuthority	Provided(p.DeliveryDetails, ta)	<>Bond(p, sf) WPassed(p, ta.Requirements)
Achieve	RequirementsMet	Package, SortFacility TransportAuthority	Bond(p, sf)	<>Passed(p, ta.Requirements)
Achieve	ClearanceProvided	Package, TransportAuthority	Provided(p.DeliveryDetails, ta) ^ Passed(p, ta.Requirements	<>Cleared(p, ta)
Achieve	DeliveryFulfillment	Package, SortFacility, Customer	Arrives(p, sf)	<>Delivered(p, c)
Achieve	SortedToDestination	Package, SortFacility	Arrives(p, sf)	<>Sorted(p, p.Destination)
Achieve	DeliveryDetailsKnown	Package, SortFacility	Arrives(p, sf)	<>Known(p.DeliveryDetails, sf)
Achieve	PackageAtDestinationRoute	Package, TransportAuthority	Known(p.DeliveryDetails, sf)	<>Sorted(p, p.Destination)
Achieve	ScannedAtNewLocation	Package, SortFacility	Arrives(p, sf)	<>Scanned(p, sf)
Achieve	ScannedAtDestinationRoute	Package, SortFacility	Sorted(p, p.Destination)	<>Scanned(p, sf)
Achieve	ScannedWhenBond	Package, SortFacility, TransportAuthority	Bond(p, sf)	<>Scanned(p, sf)

Fig. 4. Definitions for Traceable 'Package Sorting' Goals

6.2   Applying GoalBPM

We apply the proposed GoalBPM to prove the satisfaction relationship between the goal model and the process model. We specify effects as they would be defined formally due to space limitations (e.g. *Arrives(p, sf)* is equivalent to *PackageArrivesAtTheSortFacility*, as can be inferred from the goal definitions in Figure 4).

Analyzing Traceability. We firstly declare boundaries to help guide an evaluation of the traceability relation by identifying the required *pre* and possible *post* conditions for the 'Package Sorting' process. *Pre* conditions include:

#	Activity	Effect Annotation
	Pre Conditions	Arrives(p, sf) AND Provided(p.DeliveryDetails, ta)
	Possible Effects	Passed(p, ta.Requirements) AND Cleared(p, ta)
1	Scan Package	Known(p.DeliveryDetails, sf) AND Scanned(p, sf)
2	Bond Package	Bond(p, ta) AND Scanned(p, sf)
		Bond(p, ta) AND Scanned(p, sf) AND Held(p, ta)
3	Release Package	Passed(p, ta.Requirements) AND Cleared(p, ta)
4	Sort to Destination	Sorted(p, p.Destination)
5	Apply Tracking Scans	Scanned(p, sf)

T#	A#	Cumulative Effect Assessment
1	0	Arrives(p, sf) AND Provided(p.DeliveryDetails, ta) AND Passed(p, ta.Requirements) AND Cleared(p, ta)
	1	Known(p.DeliveryDetails, sf) AND Scanned(p, sf)
	4	Sorted(p, p.Destination)
	5	Scanned(p, sf)
2	0	Arrives(p, sf) AND Provided(p.DeliveryDetails, ta)
	1	Known(p.DeliveryDetails, sf) AND Scanned(p, sf)
	2	Bond(p, ta) AND Scanned(p, sf) AND Held(p, ta)
3	0	Arrives(p, sf) AND Provided(p.DeliveryDetails, ta)
	1	Known(p.DeliveryDetails, sf) AND Scanned(p, sf)
	2	Bond(p, ta) AND Scanned(p, sf)
	3	Passed(p, ta.Requirements) AND Cleared(p, ta)
	4	Sorted(p, p.Destination)
	5	Scanned(p, sf)

T#	A#	Goal Satisfaction
1	Pre	InformationProvided
		PackageCleared
		ClearanceProvided
	1	DeliveryDetailsKnown
		ScannedAtNewLocation
	4	PackageAtDestinationRoute
		SortedToDestination
	5	ScannedAtDestinationRoute
2	Pre	InformationProvided
	1	DeliveryDetailsKnown
		ScannedAtNewLocation
	2	PackageHeld
		ScannedWhenBond
3	Pre	InformationProvided
	1	DeliveryDetailsKnown
		ScannedAtNewLocation
	2	ScannedWhenBond
		PackageCleared
	3	RequirementsMet
		ClearanceProvided
	4	SortedToDestination
		PackageAtDestinationRoute
	5	ScannedAtDestinationRoute

Fig. 5. Tabulated Effect Annotation, Trajectory Decomposition, and Goal Satisfaction

Arrives(p, sf) AND Provided(p.DeliveryDetails, ta). *Post* conditions include: Sorted(p, p.Destination) OR Held(p, ta).

The initial pass at goal traceability for the 'Package Sorting' process identified three high-level goals. Further analysis identifies the specific refinements that are required for satisfaction at some point during process enactment. These goals are declared in Figure 3 and defined in Figure 4.

Effect Annotation. Firstly, we annotate the model with effects to identify the achievable (and alternative) outcomes of activities in the current process. We also include the pre-conditions themselves, and any other relevant/influential effects that may be caused by other processes that have a direct impact on process decisions and coordination. These annotations are listed in Figure 5.

Process initiation is governed by two conditions: the *arrival of packages to the sort facility* and the *provision of package information to transport authorities*. It is also identified that the prior provision of information to authorities may allow for the rapid *clearance of packages for delivery* prior to the sorting process initiating. This may occur if the requirements of the authority can be identified as being met. These effects are also added to the list of relevant/influential effects that may have occurred prior to process initiation.

Each activity is then analyzed and annotated with normal and exceptional effects. *Scanning a package* results in the *delivery details being known*. The package may be *bonded for clearance* with another *scan* being applied, and alternatively *held* if the transport authority requests. The latter is an exceptional effect that occurs due to the package meeting some characteristics that require it to be given to the authorities that then take sole control of the package from that point on. The outcome of releasing a package is the *passing of transport author-ity requirements* and ultimately *clearance*. The sorting activity then results in the *package being sorted to its destination*. The final *scanning* activity results in another update of package location.

Trajectory Decomposition. We now decompose trajectories. An analysis of the conditions influencing the choice of paths at decision gateways also guides the choice of prior and/or alternative effects when accumulating them for any given trajectory. We identify that that there are three trajectories represented in the process model. These are listed in Figure 5 with the actions that compose the trajectory, and the effects chosen from associated annotations.

The first trajectory decomposed from the model represents the *prior clear-ance of a package*, resulting in the eventual *sorting of the package* to its des-tination. The second trajectory results in packages requiring *bonding prior to clearance*, however the exceptional alternative is selected based on the adjacent decision gateway, and the package is *held by the transport authority*. The final trajectory is categorized by *requiring the package to be bonded* however, the pack-age is eventually *cleared by passing the authorities requirements*, as well as being *sorted* to, and *scanned at its destination*.

Satisfaction Analysis. We iterate through each trajectory, accumulate effects, and correlate accumulations with the desired effects and their temporal ordering on the goal model. Goals must be satisfied by firstly achievement of their pre conditions (i.e. the antecedent) with the subsequent achievement of the conse-quent in conformance to the temporal pattern chosen. Goal definitions are listed in Figure 4, with the results of the satisfaction process listed in Figure 5.

The satisfaction process identifies two normal trajectories and one excep-tional trajectory representing a *weak* satisfaction relationship between the goal model and the process model. The achievement of the operational objective to sort packages to their destinations occurs in trajectories (1) and (3), however it is not achieved in (2) due to an exceptional result occurring when the package is bonded (i.e. it is held).

6.3   Changes to the Goal Model

We now introduce the newly acquired regulatory requirements for package screen-ing that were mentioned previously. These requirements are added to the re-quired goals for satisfaction by the process.

Goal Alterations. Required alterations to the goal model, whether they be the addition, removal or modification of goals, ultimately result in modification to the *desirability* and/or *temporal ordering* of effects. Take for example, the newly acquired requirement for package screening, as represented on the adjusted portion of the goal model in Figure 6. This has introduced a newly desired effect requiring *all packages be screened by the transport authority once they arrive at a sort facility*, or formally:

∀p: Package, sf: SortFacility, ta: TransportAuthority
Arrives(p, sf) ⇒ ◊Screened(p, ta)

Additions of new effects to goal models frequently impact other goals and subsequently their refinements. As noted on Figure 6, the desired *Screened(p, ta)* is also added to the pre conditions for *PackageHeld*, as well as the *PackageCleared* goal and its refinements.

Process Implications and Evolution. Any alterations to the goal model will proportionately affect the desired *achievement* or *coordination* of effects within the business processes that are assigned their operationalization. We re-evaluate the satisfaction relationship between the 'Package Sorting' process and its goals, and apply some informal analysis to identify specific changes required at the process level.

Upon evaluation of the satisfaction relationship, it is identified that previously normal trajectories (i.e. (1) and (3)), are now also exceptional due to their inability to satisfy regulatory requirements. This is consistent with the modifications to the goal model.

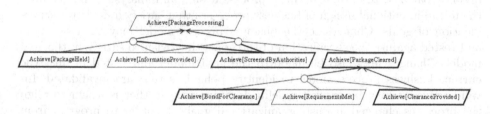

Fig. 6. Goal Model Additions for Screening Requirements

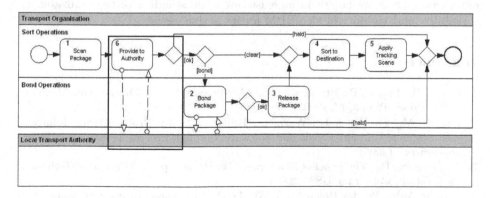

Fig. 7. Process Evolution to Achieve Goal Satisfaction

Upon further analysis of unsatisfied goals, we deduce that the *Screened(p, ta)* effect is required at some time after package arrival and prior to package clearance including bonding, as well as package holding. A decision is made as to the addition of the desired screening outcome within the process as an activity shown in Figure 7.

The process model is then re-evaluated. The included activity (6) is annotated with a normal effect that realizes *Screened(p, ta)*, and the exceptional effect of *Held(p, ta)* is also identified as possibly resulting from the activity. A decision gateway is applied adjacent to the screening activity to evaluate the outcome of the activity, and redirect process flow. Upon trajectory decomposition a new exceptional trajectory is decomposed resulting from the prior addition of the decision gateway. Goal satisfaction is then re-evaluated. The changes to the process are successful at resolving the unsatisfied relationship by achieving a weak satisfaction relationship with two normal and two exceptional trajectories.

7   Conclusion

We have proposed the GoalBPM methodology that can be used to identify the satisfaction of a process model against a goal model. The example we presented, provides a brief and informal overview. There are many possible benefits in applying GoalBPM to current business process design and analysis. This includes the initial intentional design of business processes that satisfy a deliberate specification of goals. Changes to the business process model may then be made and tested against the specification of the goals they wish to satisfy in the goal model. Changes may also be made to the goal model and tested against the current business process model to identify behaviors that are invalidated. Invalid behavior may be explicitly defined, supporting further redesign to align the processes changed against organizational goals. In order to progress from the current state, the need for formalism, tool support and testing against a large, non-trivial business case is required. We are actively pursuing these requirements, which we hope will increase our understanding of the realizability, workability and viability of GoalBPM for its active use.

References

1. Smith, H., Fingar, P.: Business Process Management: The Third Wave. Meghan-Kiffer Press, Tampa, FL (2003)
2. Dumas, M., van der Aalst, W.M., ter Hofstede, A.H.: Process-Aware Information Systems: Bridging People and Software Through Process Technology. Wiley-Interscience (2005)
3. McGoveran, D.: The benefits of a bpms. Technical report, Alternative Technologies, Felton, California, USA (2002)
4. van der Aalst, W., ter Hofstede, A., Weske, M.: Business process management: A survey. In: BPM'03 - International Conference on Business Process Management, Berlin, Springer-Verlag, Lecture Notes in Computer Science (2003) 1–12

5. Youngblood, M.D.: Winning cultures for the new economy. Strategy and Leadership **28**(6) (2000) 4–9
6. Kavakli, E.: Modelling organizational goals: Analysis of current methods. In: Proceedings of the 2004 ACM Symposium on Applied Computing, Nicosia, CY (2004) 1339–1343
7. Pyke, J., Whitehead, R.: Do better maths lead to better business processes? Business Process Trends, http://www.bptrends.com (2004)
8. Wynn, D., Eckert, C., Clarkson, P.J.: Planning business processes in product development organisation s. In: REBPS'03 - Workshop on Requirements Engineering for Busines s Process Support, Klagenfurt/Veldern, Austria (2003)
9. Watkins, R., Neal, M.: Why and how of requirements tracing. IEEE Software **11**(4) (1994) 104–106
10. van Lamsweerde, A.: Goal-oriented requirements engineering: A guided tour. In: RE'01 - International Joint Conference on Requirements Engi neering, Toronto, IEEE (2001) 249–263
11. White, S.: Business Process Modeling Notation (BPMN), Version 1.0. Business Process Management Initiative (BPMI.org). 1.0 edn. (2004)
12. Letier, E., van Lamsweerde, A.: Deriving operational software. In: FSEí10 - 10th ACM SIGSOFT Symp. on the Foundations of S oftware Engineering. (2002)
13. Letier, E.: Reasoning about Agents in Goal-Oriented Requirements Engineerin g. PhD thesis, Universite Catholique de Louvain, Louvain, Belgium (2001)
14. van Lamsweerde, A., Letier, E.: Handling obstacles in goal-oriented requirements engineering. IEEE Transactions on Software Engineering **26**(10) (2000) 978–1005
15. Bider, I., Johannesson, P.: Tutorial on: Modeling dynamics of business processes – key for building next generation of business information systems. in: The 21st international con-ference on conceptual modeling (er2002), tampere, fl, october 7-11, 2002. In: 21st International Conference on Conceptual Modeling (ER2002), Tampere, FL (2002)
16. Khomyakov, M., Bider, I.: Achieving workflow flexibility through taming the chaos. In: OOIS'00 - 6th International Conference on Object Oriented I nformation Systems, Springer-Verlag, Berlin (2000) 85–92
17. Andersson, T., Andersson-Ceder, A., Bider, I.: State flow as a way of analysing business processes - case stud ies. Logistics Information Management **15**(1) (2002) 34–45

Heuristic and Rule-Based Knowledge Acquisition: Classification of Numeral Strings in Text

Kyongho Min[1], Stephen MacDonell[1], and Yoo-Jin Moon[2]

[1] School of Computer and Information Sciences
Auckland University of Technology, New Zealand
{kyongho.min, stephen.macdonell}@aut.ac.nz
[2] Department of Management Information Systems
Hankook University of Foreign Studies, Korea
yjmoon@hufs.ac.kr

Abstract. This paper describes the rule-based classification of numerals and strings that include numerals, composed of a number and semantic unit(s) that indicate a SPEED, NUMBER, or other measure, at three levels: morphological, syntactic, and semantic. The approach employs three interpretation processes: word trigram construction with tokeniser, rule-based processing of number strings, and n-gram based classification. We extracted numeral strings from 378 online newspaper articles, finding that, on average, they comprised about 2.2% of the words in the articles. To manually extract n-gram rules to disambiguate the number strings' meanings, our approach was trained on 886 numeral strings and tested on the remaining 3251 strings. We implemented two heuristic disambiguation methods based on each category's frequency statistics collected from the sample data, and precision ratios of both methods were 86.8% and 86.3% respectively. This paper focuses on the acquisition and performance of different types of rules applied to numeral strings classification.

1 Introduction

Most efforts directed towards understanding natural language in text focus on sequences of alphabetical character strings. However, the text may include different types of data such as numeric (e.g. "25 players") - or alpha-numeric (e.g. "25km/h") - with/without special symbols (e.g. "$2.5 million") [5]. In current natural language processing (NLP) systems, such strings are treated as either a numeral (e.g. "25 players") or as a named entity (NE e.g. "$2.5 million") at the lexical level. However, ambiguity of semantic/syntactic interpretation can arise for such strings at the lexical level only: for example, the number "21" in the phrase "he turns 21 today" can on the surface be interpreted as any of the following: (a) as a numeral of NP (noun phrase) – indicating NUMBER; (b) as a numeral of NP – indicating the DAY of a date expression; or (c) as a numeral of NP – indicating AGE at the lexical meaning level. This type of numeral string is called a *separate numeral string* (e.g. the quantity in "survey of *801* voters") in this paper. Some numeral strings would not be ambiguous because of their meaningful units, and they are referred to as *affixed numeral strings* (e.g. speed in "his serve of *240km/h*").

A. Hoffmann et al. (Eds.): PKAW 2006, LNAI 4303, pp. 40–50, 2006.

In the case of separate numeral strings, some structural patterns (e.g. DATE) or syntactic functional relationships (e.g. QUANTITY as either a modifier or a head noun) could be useful in their interpretation. However, affixed numeral strings require the understanding of some meaningful units such as SPEED ("km/h" in "250km/h"), LENGTH ("m" in "a 10m yacht"), and DAY_TIME ("am", "pm" in "9:30pm").

Past research has rarely studied the understanding of varieties of numeral strings. Semantic categories have been used for named entity recognition (e.g. date, time, money, percent etc.) [7] and for a Chinese semantic classification system [13]. Semantic tags (e.g. date, money, percent, and time) and a character tokeniser to identify semantic units [1] were applied to interpret limited types of numeral strings. Numeral classifiers to interpret money and temperature in Japanese [11] have also been studied. The ICE-GB grammar [8] treated numerals as one of cardinal, ordinal, fraction, hyphenated, multiplier with two number features - singular and plural.

Polanyi and van den Berg [9] studied anaphoric resolution of quantifiers and cardinals and employed quantifier logic framework. Zhou and Su [14] employed an HMM-based chunk tagger to recognise and classify names, times, and numerical quantities with 11 surface sub-features and 4 semantic features like FourDigitNum (e.g. 1990) as a year form, and SuffixTime (e.g. a.m.) as a time suffix (see also [3] and [10] for time phrases in weather forecasts). FACILE [2] in MUC used a rule-based named entity recognition system incorporating a chart-parsing technique and semantic categories such as PERSON, ORGANISATION, DATE, and TIME.

We have implemented a numeral interpretation system that incorporates word trigram construction using a tokeniser, rule-based processing of number strings, and n-gram based disambiguation of classification (e.g. a word trigram - left and right strings of a numeral string). The rule-based number processing system analyses each number string morpho-syntactically in terms of its type. In the case of a separate numeral string, its assumed categories are produced at the lexical level. For example, "20" would be QUANT, DAY, or NUMBER at the lexical level. However, affixed numeral strings require rule-based processing based on morphological analysis because the string has its own meaningful semantic affixes (e.g. speed unit in "24km/h"). In this paper, the different types of rule needed to classify numeral strings are described in detail.

In the next section, the categories and rules used in this system are described. In section 3, we describe the understanding process for both separate and affixed numeral strings in more detail, and focus on classification rules. Section 4 describes preliminary experimental results obtained with this approach, and discussion and conclusions follow.

2 Syntactic-Semantic Categories and Rules

In this section, semantic and syntactic categories and rules (i.e. context-free rules used for affixed numeral strings) used to parse numeral strings in real text are described.

This system uses both syntactic and semantic categories to understand separate and affixed numeral strings, because a numeral string such as "20" (i.e. separate numeral string) can be understood by itself (as in "20 pages") or with reference to a structural relationship to adjacent strings (as in "on September 20 2003"). The separate numeral string "20" in "20 pages" can be interpreted as a QUANTITY to modify the noun "pages". However, knowledge of the specific DATE representation (structural relationships between adjacent strings) in "on September 20 2003" is needed to understand "20" as DAY. This requirement is even more evident with "7/12/2003" which can mean July 12, 2003 (US) or 7 December, 2003 (e.g. in Australia and New Zealand). Thus semantic categories including DAY, MONTH, and YEAR are used for date representation.

We use 40 syntactic and semantic categories, including specific semantic categories for some numeral strings (e.g. semantic categories (e.g. MONEY, DATE) and syntactic categories – (e.g. NUMBER, FLOATNUMBER, FMNUMBER) – Table 1). For example, the category FMNUMBER (ForMatted Number) signals numbers that frequently include commas every 3 digits to the left of the unit digit for ease of reading, as in "*5,000* peacekeepers."

Table 1. Sample categories and their examples

Category	Example in Real World Text
Age	"mature *20-year-old* contender", "he turns *21* today"
Date	"*20.08.2003*"
Day	"August *11* 2005"
Daytime	"between *9:30am* and *2am*", "at *3* o'clock"
Floatnumber	"support at *26.8* per cent"
FMnumber	"took command of *5,000* peacekeepers"
Length	"a *10m* yacht"
Money	"spend *US$1.4* billion"
Name	"Brent crude *LCOc1*"
Number	"*8000* of the Asian plants"
Ordinal	"a cake for her *18th* birthday"
Plural	"putting a *43-man* squad"
Phone-Number	"ph: (09) 917 1234"
Quant	"survey of *801* voters"
Range	"for *20-30* minutes"
Scores	"a narrow *3-6* away loss to Otago"
Speed	"His serve of *240km/h* this season"
Street-Number	"Address: *123* Moutain rd Mt. Eden"
Temperature	"temperatures still above *40C*"
Year	"by September *2026*"

There are two types of dictionaries in our system: one for normal English words with syntactic information such as lexical category, number, and verb's inflectional form. The other dictionary (called the user-defined dictionary) includes symbol tokens (e.g. "(", ")") and units (e.g. "km", "m"). For example, the lexical information for "km" is (:POS (Part of Speech) LU (Length Unit)) with its meaning KILOMETER.

The system uses 64 context-free rules to represent the structural form of affixed numeral strings. Each rule describes relationships between syntactic/semantic categories of the components (e.g. a character or a few characters and a number) produced by morphological analysis of the affixed string. Each rule is composed of a LHS (left hand side), RHS (right hand side), and constraints on the RHS (e.g. DATE → (DAY

DOT MONTH DOT YEAR), Constraints: ((LEAPDATEP DAY MONTH YEAR))).
The interpretation rules are discussed in the next section in detail.

3 Numeral String Classification

The numeral string interpretation algorithm is composed of three processes: a morphological analysis module, a rule-based interpretation module, called ENUMS (English NUMber understanding System), which employs both a CFG (Context-Free Grammar) augmented by constraints and a parser, and a category disambiguation module to select the best category of an ambiguous numeral string by using word trigrams.

3.1 Morphological Analysis of Numeral Strings

Affixed numeral strings such as "240km/h serve" and "a 10m yacht" require knowledge of their expression formats (e.g. speed → number + distance-unit + slash + time-unit) for understanding. For example, the string "240km/h" is analysed morphologically into "240" + "km" + "/" + "h". Our morphological analyser considers embedded punctuation and special symbols. In the case of the string "45-year-old", the morphological analyser separates it into "45" + "-" + "year" + "-" + "old". Thus we use the term, morphological analysis, rather than tokenisation because each analysed symbol is meaningful in numeral string interpretation. Table 2 shows some more results from the morphological analyser.

Table 2. Examples of morphological analysis of numeral strings

Category	Example	Morphological Analysis
MONEY	"($12.56)"	"(" + "$" + "12" + "." + "56" + ")"
DATE	"20.08.2003"	"20" + "." + "08" + "." + "2003"
FMNUMBER	"2,000"	"2" + "," + "000"
SPEED	"240km/h"	"240" + "km" + "/" + "h"
RANGE/SCORES	"20-30"	"20" + "-" + "30"
DAYTIME	"9:30am"	"9" + ":" + "30" + "am"
FLOATNUMBER	"0.03"	"0" + "." + "03"
CAPACITY	"8.2μmol/L"	"8.2" ("8" + "." + "2") + "μmol" + "/" + "L"
PLURAL	"1980s"	"1980" + "s"

After analysing the string, dictionary lookup and a rule-based numeral processing system based on a simple bottom-up chart parsing technique [6] are invoked. Instances that include some special forms of number (e.g. "03" in a time, day), are not stored in the lexicon. Thus if the substring is composed of all digits, then the substring is assigned to several possible numeric lexical categories. For example, if a numeral string "03" is encountered, then the string is assigned to SECOND, MINUTE, HOUR, DAY, MONTH, and BLDNUMBER (signifying digits after a decimal point, e.g. "0.03"). If the numeral string is "13" or higher, then the category cannot be MONTH. Similar rules can be applied to DAY and other categories. However, "13" can clearly be used as a quantifier.

Non-numeral strings are processed by dictionary lookup as mentioned above, and their lexical categories used are necessarily more semantic than in regular parsing. For example, the string "m" has three lexical categories: LU (Length Unit) as a METER (e.g. "a 10m yacht"), MILLION (e.g. "$1.5m"), and TU (Time Unit) as a MINUTE (e.g. "12m 10s" - 12 minutes and 10 seconds).

After morphological processing of substrings, an agenda-based simple bottom-up chart parsing process is applied with 64 context-free rules that are augmented by constraints. If a rule has a constraint, then the constraint is applied when an (inactive) phrasal constituent is created. For example, the rule to process a date of the form "28.03.2003" is DATE → (DAY DOT MONTH DOT YEAR) with the constraint (LEAPYEARP DAY MONTH YEAR), which checks whether the date is valid. An inactive phrasal constituent DATE1 with its RHS, (DAY1 DOT1 MONTH1 DOT2 YEAR1), would be produced and the constraint applied to verify the well-formedness of the inactive constituent.

The well-formedness of DATE (e.g. "08.12.2003") is verified by evaluating the constraint (LEAPDATEP DAY MONTH YEAR). Some other rules for affixed/separate numeral string interpretation are:

RULE5   **LHS:** AGE
         RHS: (NUMBER HYPHEN NOUN HYPHEN AGETAG) – e.g. "*38-year-old* man"
         Constraints: ((INTEGER-NUMBER-P NUMBER) (SEMANTIC-AGE-P NOUN) (SINGULAR-NOUN-P NOUN))

RULE21   **LHS:** TEMPERATURE
         RHS: (NUMBER CELC) – e.g. "*40C*"
         Constraints: (INTEGER-NUMBER-P NUMBER)
         Where CELC means CELsius-C

RULE22   **LHS:** RANGE
         RHS: (NUMBER HYPHEN NUMBER) – e.g. "*20-30* minutes"
         Constraints: (RANGE-P NUMBER NUMBER)

RULE30   **LHS:** FLOATNUMBER
         RHS: (NUMBER DP BLDNUMBER) – e.g. "*20.54* percent"
         Constraints: (FLOATNUMBER-P NUMBER DP BLDNUMBER)
         where DP means Decimal Point and BLDNUMBER means BeLow-Decimal NUMBER.

RULE42   **LHS:** WEIGHT
         RHS: (NUMBER WU) – e.g. "55kg"
         Constraints: NIL.
         where WU means Weight Unit.

3.2 Classification Based on Word Trigrams

For separate numeral strings, the interpreted categories can be ambiguous because there is no semantic unit attached. For example, "240km/h" would be uniquely interpreted as SPEED. However, the numeral string "20" could be either QUANT (e.g. "20 boys") or DAY (e.g. "20 May 2005") without using different context information. Thus word trigrams are used to disambiguate the syntactic/semantic categories of numeral strings.

Word trigrams are collected when a document is read and tokenised. While tokenising a string (tokenisation based on a single whitespace), the numeral string is identified with its word trigram (left and right string of the numeral string). For example, the numeral string "100" in "The company counts more than 100 million registered users worldwide." has its word trigram ("than" - left wordgram, "million" - right wordgram). If a numeral string occurs at either the start or end of a sentence, then either a left or right wordgram would be empty (i.e. NULL).

Table 3. Examples of feature types

Feature Type	Examples
Lexical Cateogry	Preposition-p: (preposition-p left wordgram ("to")) in "home to 22 superyachts"
Number information	Plural-noun-p: (plural-noun-p right wordgram ("voters")) in "801 voters"
Validity of value	Valid-day-p: (valid-day-p numeral string ("28") right wordgram ("February")) in "28 February 2004" – not "30 February 2004"
Conceptual type	Month-string-p: (month-string-p right wordgram ("February")) in "28 February 2004"
Case of a word	Capital-letter-p: (capital-letter-p left wordgram ("Lee,")) in "Lee, 41, has…"
Punctuation marks	Comma-p: (comma-p left wordgram ("Lee,")) in "Lee, 41, has…"

Disambiguation of categories is based on rules manually encoded by using sample data and each rule is based on morpho-syntactic features of the word trigrams. For example, a punctuation mark like comma is important for disambiguating the AGE category (e.g. "41" in "Lee, 41, has").

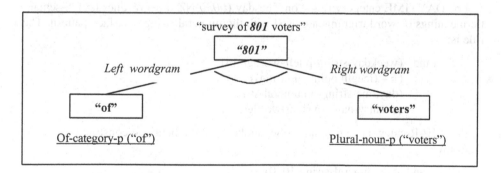

Fig. 1. Contextual constraints based on word trigrams

The features used for selection of the best meaning of an interpreted numeral string are based on syntactic and surface features. These features are joined together to reflect contextual information (e.g. using neighbour words information). The current system uses left and right adjacent words of the numeral string with the following features (Table 3): lexical category (e.g. NOUN, VERB), number information (e.g. PLURAL, SINGULAR), validity of values (e.g. valid DAY), semantic information (e.g. MONTH concept), Case of a letter (e.g. capitalisation), and punctuation marks (e.g. PERIOD, COMMA).

With the features extracted from a numeral string's word trigram, the contextual features of the word trigram are used for the selection of the 'best' category for that numeral string. The contextual information is extracted manually and its form is based on conjunction of the word trigrams features (Fig. 1).

For QUANT category disambiguation, 22 constraints are used. The word trigram (wordgram) for a numeral string "801" in the substring "survey of 801 voters" would be ("of" "801" "voters"). Thus one QUANT selection rule would be:

> (and (of-category-p left-wordgram ("of"))
> (plural-noun-p right-wordgram ("voters"))).

If the numeral string "20" is in the string "March 20 2003", then the category would be DAY and one of four selection rules would be:

> (and (month-string-p left-wordgram ("March"))
> (valid-day-p "March" "20") - not "March 35"
> (number-p right-wordgram ("2003"))).

If a numeral string (e.g. "41" in "Lee, 41, has") satisfies one of three AGE rules, then the numeral string is disambiguated as AGE. The rule is:

> (and (capital-letter-p left-wordgram("Lee"))
> (comma-p left-wordgram ("," in "Lee,"))
> (comma-p numeral-string ("," in "41,"))).

This rule means that if the word in the left wordgram begins with a capital letter, if the word in the left wordgram has a comma, and if the numeral string has a comma, then the rule applies.

For DAYTIME category (e.g. "on Tuesday (*0030* NZ time Wednesday)"), semantic meanings of word trigrams are used with the numeral string's surface pattern. The rule is:

> (and (weekday-string-p left-wordgram)
> (= 4 (length numeral-string))
> (daytime-string-p numeral-string)
> (country-name-p right-wordgram)).

For YEAR category (e.g. "end by September *2026*"), heuristic constraints are used as follows:

> (and (>= numeral-string 1000)
> (<= numeral-string 2200)).

The contextual information based on word trigrams is applied to disambiguate multiple categories resulting from the numeral string interpretation process. Two heuristic methods are implemented to compare their results:

- Heuristic Method 1 (Method-1) – Method 1 applies wordgram constraints and collects and then considers all satisfied constraints. For example, if the QUANT and NUMBER constraints for a numeral string are satisfied, then the two categories are used for the numeral's disambiguation. With collected categories, the annotation frequency of the categories collected from sample data (i.e. Sample data in Table 4) is used to select the best category. If the frequency of QUANT is greater than that of NUMBER in annotation statistics, then QUANT is selected for the category of the numeral string from the meanings processed by a numeral string interpretation system. If a constraint for AGE category is satisfied, then a new category, AGE, is produced because the numeral string interpretation system could not produce the category. If there is no category that satisfies the constraints, then preference rules with ordered categories based on frequency of annotation (e.g. QUANT > MONEY > DATE > etc.) are applied to select the best category.

- Heuristic Method 2 (Method-2) – Method 2 is similar to Method-1 except for the application of annotation statistics. To select the best category for an ambiguous numeral string, the rareness of annotation statistics is applied. If both YEAR and QUANT constraints are satisfied, and the annotation frequency of YEAR is less that that of QUANT, then YEAR is selected for the category of the numeral string. If there is no category that satisfies the constraints, then preference rules with ordered categories based on rareness of annotation statistics (e.g. DATE < MONEY < YEAR < etc.) are applied to select a category.

4 Experimental Results

We implemented our system in Allegro common lisp with IDE. We collected 9 sets of online newspaper articles and used 91 articles (sample data) to build disambiguation rules for the categories of numeral strings. The remaining 287 articles (test data) were used to test the system. Among the 48498 words in the 91 sets of sample data, 886 numeral strings (1.8% of total strings and 10 numeral strings out of 533 strings for each article on average) were found. In the case of the test data, 3251 out of 144030 words (2.2% of total strings and 11 numeral strings of 502 strings for each article on average) were identified as numeral strings (Table 4).

Table 4. Data size and proportion of numeral strings

Date Name	Total Articles	Total Strings	Total Numerals	Remark
Sample	91	48498	886 (1.8%)	2 sets
Test	287	144030	3251 (2.3%)	7 sets
Total	378	192528	4137 (2.1%)	-

The proportion of numeral strings belonging to each category in both sample and test data were QUANT (826 of 3251, 20.0%, e.g. "survey of *801* voters"), MONEY (727, 17.6%, e.g. "$15m", "$2.55"), DATE (380, 9.2%, e.g. "02.12.2003"), YEAR (378, 9.1%, e.g. "in 2003"), NUMBER (300, 7.3%, e.g. "300 of the Asian plants"), SCORES (224, 5.4%, e.g. "won 25 - 11"), FLOATNUMBER (8.0%, e.g. "12.5 per cent"), and others in order.

Table 5 shows the recall/precision/F-measure ratios (balanced F-measurement) based on the two disambiguation methods. Method-2 for the test data shows better recall ratio (77.6%) than Method-1. Method-1 for the test data shows better precision ratio (86.8%) than Method-2. However, the difference between Method-1 and Method-2 is 0.5% in recall ratio and 0.5% in precision ratio, indicating that the performance of each method is close to identical.

Table 5. Recall/Precision/F-measurement ratios of two heuristic methods by data set

DateName	Recall Ratio (%)		Precision Ratio (%)		F-measure (%)	
	Method-1	Method-2	Method-1	Method-2	Method-1	Method-2
Sample	86.0	86.6	95.3	93.4	90.2	89.8
Test	77.1	77.6	86.8	86.3	81.7	81.7
Average	81.5	82.1	91.1	89.8	85.9	85.8

Table 6 shows the recall/precision/F-measure ratios, of selected categories, based on the two disambiguation methods. The average performance difference between the methods is 0.7% in recall ratio, 0.6% in precision ratio, and 0.1% in F-measurement. The results for affixed numeral strings (e.g. FLOATNUMBER, FMNUMBER, MONEY, RANGE, SCORES) showed large differences in performance (10.3% to 100%) as did the results for separate numeral strings (e.g. AGE, DAY, NUMBER, QUANT, YEAR) (38.2% to 97.5%). The performance of the RANGE category is poor because of numeral strings such as "$US5m-$US7m" and "10am-8pm". These numeral strings are presently not covered by our CFG (Context-Free Grammar).

Table 6. Recall/Precision/F-measurement ratios of two heuristic methods by categories

Category	Recall Ratio (%)		Precision Ratio (%)		F-measure (%)	
	Method-1	Method-2	Method-1	Method-2	Method-1	Method-2
Age	43.8	43.8	97.5	97.5	60.5	60.5
Day	68.7	69.3	86.8	81.9	76.7	75.1
Number	82.1	91.7	42.4	38.2	55.9	53.9
Quant	83.2	81.3	88.2	97.5	85.6	88.7
Year	86.9	70.0	92.6	87.9	89.6	77.9
Floatnumber	98.6	98.6	96.9	96.9	97.8	97.8
Fmnumber	100	100	98.8	98.8	99.4	99.4
Money	99.2	99.2	99.9	99.9	99.5	99.5
Range	10.3	55.2	66.7	35.6	17.9	43.2
Scores	84.8	74.1	89.2	97.1	87.0	84.1
*Average	74.9	75.6	90.0	89.4	81.8	81.9

(*average is the average of all categories)

Compared to other separate categories, the system interprets the QUANT category better than YEAR and NUMBER because the disambiguation module for QUANT category has more constraints based on wordgram information (i.e. 22 constraints for QUANT, 7 for YEAR, and 6 for NUMBER).

For disambiguation process using Method-1 and Method-2, 2213 (53.5%) of 4137 numeral strings were ambiguous after numeral string interpretation process. Among these, the numbers of satisfied constraints are no constraint (746 – 33.7%), one constraint (1271 – 57.4%), 2 constraints (185 – 8.4%), and three constraints (11 – 0.5%).

5 Discussion and Conclusions

It is not easy to compare our system to other NE (Named Entity) recognition systems directly because the target recognition of named entities is different. Other systems in MUC-7 [2] and CoNLL2003 [4] focused on the general recognition task of named entities including person, location, date, money, and organisation. The systems in MUC-7 and CoNLL2003 were trained and tuned by using the necessary training corpus with document preprocessing (e.g. tagging and machine learning). However, our system is focused on understanding the varieties of numeral strings more deeply and had no training phase. The manually annotated data from sample data (25%) was used for implementation of disambiguation rules. Performance of MUC-7 systems was greater than 90% in precision. Our system correctly interpreted 88.7% of numeral strings.

Further rules and lexical information are required to process more numerals in real world text (see [5]). For better disambiguation, more fine-grained disambiguation modules based on word trigrams would be required. Currently, syntactic categories of both left and right wordgrams of each numeral string are used. The major problem of the use of syntactic category is lexical ambiguity (e.g. "in" is lexically ambiguous as preposition, adverb, and noun). To reduce the ambiguities, more fine-grained surface patterns would be required. In addition, the extension of this system to other data such as biomedical corpora [12] is required to test the overall effectiveness of our approach.

Another research avenue would be the automatic acquisition of constraints for the disambiguation process and the determination of the significance of each feature for the disambiguation process. For example, for the QUANT category, the plural number information of a right wordgram could be more significant than various information in a left wordgram. In addition, the significance of offset (adjacency) of wordgrams to extract more contextual knowledge could be studied for better disambiguation precision for future development.

In conclusion, separate and affixed numeral strings are frequently used in real text. However, there seems to be no system that interprets numeral strings systematically; they are frequently treated as either numerals or nominal entities. In this paper, we have analysed the numeral strings at lexical, syntactic, and semantic levels with some contextual information. The system is composed of a tokeniser with word trigram constructor, numeral string processor which includes a morphological analyser and a simple bottom-up chart parser with context-free rules augmented by constraints, and a disambiguation module based on word trigrams. The numeral string interpretation system successfully interpreted 88.7% of test data. The system could be scaled up to cover more numeral strings by extending the lexicon and rules.

References

1. Asahara, M., Matsumoto Y.: Japanese Named Entity Extraction with Redundant Morphological Analysis. Proceedings of HLT-NAACL 2003. (2003) 8-15
2. Black, W., Rinaldi, F., Mowatt, D.: FACILE: Description of the NE system used for MUC-7. Proceedings of MUC-7. (1998)

3. Chieu, L., Ng, T.: Named Entity Recognition: A Maximum Entropy Approach Using Global Information. Proceedings of the 19th COLING. (2002) 190-196
4. CoNLL-2003 Language-Independent Named Entity Recognition. http://www.cnts.uia.ac.be/conll2003/ner/2. (2003)
5. Dale, R.: A Framework for Complex Tokenisation and its Application to Newspaper Text. Proceedings of the second Australian Document Computing Symposium. (1997)
6. Earley, J.: An Efficient Context-Free Parsing Algorithm. CACM. 13(2) (1970) 94-102
7. Maynard, D., Tablan, V., Ursu, C., Cunningham, H., Wilks, Y.: Named Entity Recognition from Diverse Text Types. Proceedings of Recent Advances in NLP. (2001)
8. Nelson, G., Wallis, S., Aarts, B.: Exlporing Natural Language - working with the British Component of the International Corpus of English, John Benjamins, The Netherlands. (2002)
9. Polanyi, L., van den Berg, M.: Logical Structure and Discourse Anaphora Resolution. Proceedings of ACL99 Workshop on The Relation of Discourse/Dialogue Structure and Reference. (1999) 10-117
10. Reiter E., Sripada, S.: Learning the Meaning and Usage of Time Phrases from a parallel Text-Data Corpus. Proceedings of HLT-NAACL2003 Workshop on Learning Word Meaning from Non-Linguistic Data. (2003) 78-85
11. Siegel, M., Bender, E. M.: Efficient Deep Processing of Japanese. Proceedings of the 3rd Workshop on Asian Language Resources and International Standardization. (2002)
12. Torii, M., Kamboj, S., Vijay-Shanker, K.: An investigation of Various Information Sources for Classifying Biological Names. Proceedings of ACL2003 Workshop on Natural Language Processing in Biomedicine. (2003) 113-120
13. Wang, H., Yu, S.: The Semantic Knowledge-base of Contemporary Chinese and its Apllication in WSD. Proceedings of the Second SIGHAN Workshop on Chinese Language Processing. (2003) 112-118
14. Zhou, G., and Su, J. Named Entity Recognition using an HMM-based Chunk Tagger. Proceedings of ACL2002. (2002) 473-480

RFID Tag Based Library Marketing for Improving Patron Services

Toshiro Minami

Kyushu Institute of Information Sciences,Faculty of Management and Information
Sciences, 6-3-1 Saifu, Dazaifu, Fukuoka 818-0117, Japan
minami@kiis.ac.jp
http://www.kiis.ac.jp/~minami/

Abstract. In this paper, we deal with a method of utilizing RFID tags attached on
books and extract tips that are useful for improving library services to their patrons. RFID is an AIDC (Automatic Identification and Data Capture) technology,
with which we can automatically collect data how library materials are used and
how often. By analyzing such data we are able to acquire knowledge that helps
the librarians with better performing their jobs such as which books to collect,
how to help their patrons, and so on. We will call such method "in-the-library
marketing," because the data deal with how library materials are used by patrons
in the library. It is more effective if we also add the data captured from out-of-the-
library and integrate them for whole "library marketing." Furthermore, we also illustrate architecture for protecting the patron-related privacy data from leakage,
which is another important issue for library marketing.

Keywords: RFID (Radio Frequency Identification), IC Tag, Digital Library,
Library Automation and Digitization, Security Issue, Library Marketing.

1 Introduction

Thanks to the rapid progress of ICT (Information and Communication Technology),
we are able to access to information much easier and more rapidly than before, so that
we can use it for various problem solving in our life. As a social function to help us,
or patrons, library should change itself in order to give improved patron services
along with the change of social needs. The philosophy underlying this concept is the
one proposed by Ranganathan in his Five Laws of Library Science [10]. In this paper
we consider two important issues relating such changes: digitization and ubiquitiza-
tion. These issues are important because they contribute to acquire better knowledge
for improving patron services provided by libraries.

Relating to the first issue, it becomes popular that the announcements from a library are put also on its Web pages now. Patrons are able to know its opening hours
without giving a call and asking. OPAC (Online Public Access Catalog) service is
also provided via Web so that patrons are able to search for books they want to read
without going to the library. Some libraries provide digital images of rare materials to
public. Increasing number of e-Journals and e-Books are serviced so that patrons can

A. Hoffmann et al. (Eds.): PKAW 2006, LNAI 4303, pp. 51–63, 2006.
© Springer-Verlag Berlin Heidelberg 2006

browse the original contents at home. Such digital materials are provided on the Web based systems and thus it is easy to automatically collect data of which materials are used and which time they are used. By analyzing these data and acquire knowledge what services are more beneficial than others for patrons, librarians are able to get good tips for improved patron services.

For the second issue, the number of libraries that have installed the RFID (Radio Frequency Identification) tag system [3], which represents ubiquitous technology, is increasing rapidly in these couple of years. In Japan, for example, public libraries are more aggressive than other kinds of libraries. One of the reasons might be that many town libraries have been located in community centers and many towns have been wishing to construct library buildings of its own. It is a good chance for them to install the RFID tag system for efficiency of jobs and improvement of security and patron services.

Considering such changes, it is easy to expect that most of the library materials will be digitized and physical materials, e.g. ordinary books and magazines, will be used supplementary in the far future. Thus we are in the transitional stage from libraries of physical materials to libraries of digital materials. During this transitional period, they are hybrid libraries in which physical and digital materials are coexisting, and the ratio of digital materials is increasing gradually.

One of the most important things we have to do now is to establish a hybrid library system so that libraries can deal with both physical and digital materials in a uniform way and the transition goes seamlessly.

From this point of view, RFID technology is very appropriate for libraries to introduce. RFID is one of the technologies that are called AIDC (Automatic Identification and Data Capture). The AIDC technology gives two big advantages to libraries. First it provides a means of better and more efficient method of managing physical materials. Secondly it provides a means of automatically collect digital data how such physical materials are used. By utilizing such technologies we can easily collect data about which materials are used, when they are used, by whom they are used, and so on. The digitization in the first issue and the digital data collected with AIDC technology provides us with digital data, which are easy to be integrated and thus it is easy to construct a big and comprehensive database (DB) by collecting such whole data. In this way the data about physical materials and digital materials can be treated in a uniform way. Libraries having RFID tag system is also called "u-library (i.e. ubiquitous library)" [5, 8].

Once we have the integrated DB, we are able to extract information and acquire knowledge by applying some datamining (DM) techniques and analyze the data. In the current, i.e. traditional, libraries, the circulation data are virtually the only digital data that could be used for datamining. Thus adding up the new data collected from the digitized services and the one from RFID and acquiring useful knowledge for improving patron services are the new and challenging application field for knowledge acquisition(KA) researchers. Such knowledge is supposed to be used also for revising the ways of services and starting new library services that will be convenient for the patrons of the library. In this paper, we call such new method "in-the-library marketing" [6, 9].

Personalized patron service is very important in the next generation library services. We will call it "My Library" service. In the top page of My Library, patrons login to the service by type their library IDs and passwords. They can get their personal information: for example, which materials they are borrowing and when is the due date of them. They may get some recommended book list among the materials that are just cataloged and on the loaning service.

In order to provide such personalized services, the library needs the profile information of the patrons. The more appropriate information they have it is possible to provide more accurate and more sophisticated services to them. In this point of view, AIDC is again the very important key technology for libraries.

The AIDC technology mostly used in the libraries so far is "barcode." Comparing to the barcode technology, the RFID technology has advantages in a couple of aspects; it is faster to identify the IDs, it can read the data of tags that are located in the invisible places from the readers, e.g. tags attached inside of the books, and recognize multiple IDs, in one action. In this respect, RFID is more appropriate to capture data automatically about change of the status of materials and patrons.

This paper consists of five sections. In the next section, i.e. Section 2, we will briefly explain what the RFID tag system is like and how it is used in the libraries. In Section 3, we will discuss in what way the RFID tags are used for the in-the-library marketing, followed by the security issue in Section 4. Finally in Section 5, we conclude the discussions in this paper.

Throughout this paper, we mainly deal with how to automatically, i.e. easily, collect data that are appropriate for acquiring knowledge that are useful for improving patron services of libraries. This is the very first step of knowledge acquisition system for library marketing.

2 RFID Technology and Its Application to Libraries

In this section we will have a brief view what RFID system is all about and we put special focus on how it is currently used in library applications. First of all we describe what the RFID tag technology is like in Figure 1. The RFID system consists of two components; tags and reader/writers (R/Ws). Tags can communicate with the reader/writers which are located in near distance of the tags.

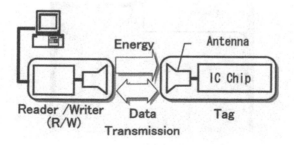

Fig. 1. RFID Tag System (Passive Type)

In Figure 1 the RFID tag at the right-hand side consists of an IC chip and an antenna. It has no batteries and thus cannot run standalone. At the left-hand side is an R/W, which provides energy to the tag. Then the tag gets energy from the R/W with electro-magnetic induction via the antenna. It waits until sufficient energy is charged. When it is ready, it communicates with R/W and exchange data such as its ID and status data by making use of the same antenna. At the backend of the R/W are applications such as databases.

The frequencies used in RFID systems range from about 100kHz up to GHz bands, which is in the ISM (Industrial, Scientific and Medical) bands [3]. The most popularly used frequency among them is 13.56MHz. It is mostly appropriate for applications with medium read distance, i.e. from about 1cm to 1m. Other frequencies such as UHF band and microwave band are also under evaluation. However, at the moment, 13.56MHz is considered to be the most appropriate one for the library application.

The tags described so far are called "passive tags" because they have no batteries and need external energy as is shown in Figure 1. There is another type of tags, which are called "active tags" because they are equipped with batteries and thus they can work without external energy supply. They will emit radio wave off the tag autonomously and the R/Ws receive the radio wave and get the data from the tag. The most important advantage of active tags is that the data transmission distance is much longer, for example about 10m to 20m, then passive tags and thus the security gate becomes more reliable. On the other hand, active tags have disadvantages such as they are thicker, heavier, much more expensive and shorter life span, probably up to a couple of years, than passive ones. However these problems might be overcome in the future and such tags may be widely used together with passive ones in libraries.

Figure 2 is an example that shows how RFID tag is attached on a book (in Chikushi Branch Library of Kyushu University Library [4], Japan). The tag is formed as a label on which the library name is marked together with the university logo. The material ID is also marked in barcode on the label. The barcode is supposed to be used when this material is carried to another library in the ILL (Inter-Library Loan), i.e. for interoperability, and when the tag has going bad and becomes broken, i.e. for insurance.

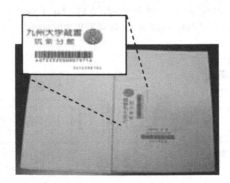

Fig. 2. RFID Tag Attached on a Book

Fig. 3. Self Checkout Machine

The tag is attached on the first leaf of the book next to the cover in this case. It is safer to be damaged than attaching on the outside of the cover. However it is more laborious to read its ID from its barcode label. You open the cover first and read the ID. On the other hand, when we read the ID by the RFID tag we just put the book near the reader without opening the cover.

As is seen in Figure 2, the tag is attached to very close place to the spine of the book. This is because it is more sensitive than attaching it at other place when we make an inventory. We use portable readers and scan the books as they are stored at the bookshelves. The reader is less powerful than the normal desktop type reader and security gate. Thus it is more preferable to arrange the tag close to the reader as possible, i.e. close to the spine.

Comparing to the barcode system which is mostly used now, RFID tag system has an advantage that it is much easier to position materials. As a result self checkout machine is easier to use so that it is good for children and elderly patrons. Figure 3 is an example. A patron is going to put a couple of books on the designated area. Then the machine will display the book IDs, the process ends when the patron pushes the OK button on the touch screen. The list of borrowed books will be printed at the right-hand side printer.

This is another type of advantage of RFID to barcode. It is not only more efficient but also more sophisticated and has much easier user-interface. This is a very important point. So far the dominating reason for the libraries whey they introduce the RFID tag system is that it is more efficient; i.e. it is faster to proceed circulation, it is supposed to have less running cost, and thus the number of librarians needed will be smaller, etc.

Here the most important motivation of introducing RFID tag system is efficiency. We will call it the step 1 of library automation. What will come next for the step 2? It should be effectiveness. By utilizing the RFID tag system we can create new methods not only for efficient but also for giving better, more sophisticated, more advanced services so that customer satisfaction increases. This is what we would like to pursue in library automation.

Fig. 4. Self Book Return Machine

Figure 4 shows another example. It is a self book return machine (in Baulkham Hills Library [2], Australia). The patrons are supposed to return their borrowing books by using this machine. Inside of the opening is a lid that is locked normally. You cannot put books inside of the machine. You are supposed to hold the book for a while as you return a book in the opening. Then the R/W installed near the lid reads the ID of the book and check if it an appropriate book for returning or not. The lock is released when the machine recognizes its appropriateness and you can drop the book into the box of the return machine. This is another good example for using RFID technology. It will be very difficult to make such a machine that discriminate the appropriate books from inappropriate ones if you use barcode for recognizing the book ID.

In a Web page of Baulkham Hills Shire Council, they say "This technology gives staff more opportunity to move from behind the desk and focus on value-added, proactive customer service," which is the most important objective of the step 2 of library automation.

3 In-the-Library Marketing with RFID

As we have pointed out in the previous sections, RFID tag system can be effectively used for automatically capturing data by many R/Ws, by which we are able to acquire knowledge that should be useful for improving library's patron services and management. Because the RFID tags and R/Ws collaboratively work and the knowledge is acquired from this collaboration, we can model this system based on multi-agent framework [8, 11].

Fig. 5. Intelligent Book Shelf

3.1 Data Collection

An intelligent shelf is a bookshelf which has R/Ws in it so that it can read which books are stored in which shelf. Note that a similar shelf used in retail store is often called smart shelf. Figure 5 is an intelligent bookshelf, in which four antennas are placed like book separators in the bookshelf. The controller activates the antennas one by one and recognizes the book IDs in the active shelf. It also changes the active bookshelves themselves one by one so that one controller can deal with tens of antennas.

Currently the bookshelf R/Ws cannot detect in which part the book is located on a shelf. However if we make this R/W system more sophisticated so that we are able to locate more accurately and eventually to locate exactly where and in what order the books are arranged on a shelf.

By using such bookshelves the library system can detect whenever a book is taken out of the shelf and whenever it is returned on a shelf. For example the library system can make a list of books that were returned on wrong shelves. By using this list librarians are able to relocate such books to their right positions.

Also such data can be used to rank the books according to the frequencies of taking out and returning, which indicate how frequently the books are used in the library. This will give a good tip to librarians when they evaluate their book collection policy.

Currently the intelligent shelf is very expensive though. One example price is one million yen, or about ten thousand US dollars, for one line of bookshelves. It is far too expensive to replace all the bookshelves currently used in libraries. However it is worth considering if we first replace just one or a couple of bookshelves with intelligent shelves and increase the number gradually.

For such purpose one good candidate is the bookshelves for newly registered books and/or for those just returned by a patron. Such books attract patrons' interest and thus they will be used in high frequencies. By analyzing such data, librarians will be helped by the extracted information with choosing new books to be purchased.

Another candidate is, specifically in university libraries, for the books that are designated as subtexts by teachers. These books usually appear in the syllabuses. It is a great benefit for students to read such textbooks in the library.

(a) A Patron Reading a Book on the Table   (b) R/Ws under the Table

Fig. 6. Intelligent Browsing Table

If we use the intelligent shelves for such books, the library can collect the detailed data how these books are used; e.g. for each book when it was taken off the shelf and when it was returned, and maybe who did it.

By collecting and analyzing these data, the library might be able to decide how many volumes of a title to buy according to the data. If a book has little or no usage history, the library can let the teacher who recommended this book know this fact Then. he/she may encourage the students of the class so that they use this textbook more.

Book trucks might be a good choice to use in some situations. In some libraries the books taken out from the ordinary bookshelves are supposed to return on book trucks. If we set shelf readers to the book trucks, the system can get the data when the books are returned to the book trucks.

An intelligent browsing table is a table in a browsing room of the library, which has R/W(s) in it. Figure 6 is an example browsing table experimented in AIREF Library in Fukuoka City, Japan [1]. In this figure a patron is reading a book on the table. He has a couple of books around him (a) and two RFID readers (b) detect them and send the data to their server.

By analyzing the data from the intelligent browsing table(s), librarians are able to obtain information which books are read, how long, how often and others. Such information is useful for shelf arrangement and book collection. If the table readers can also collect patron IDs, they can get information who reads what books. By analyzing these data we can get which and which books are often read together by such and such patrons. This information is useful for book recommendation service to patrons by use of the collaborative filtering technology [6, 12], which has been well-known in agent researcher's communities.

3.2 Analysis

Automatically or manually collected data form a big database. It consists of, for example, catalog data for materials, circulation records, data by intelligent shelves, and others. We are able to extract statistical data not only from one type of data but also from some types of data by combining them.

From circulation records, for example, we are able to know how many books a patron borrowed so far, per year, per month, and so on. We are also able to know how many books were borrowed in each day of a week, and in which time zone in a day.

By combining it with patron's personal data, we are able to know, for example, the members of a department of a university borrows what sort of materials and how many, and differences of borrowed materials between two departments.

Furthermore, we will be able to acquire knowledge something like "patrons who come to the library in the morning borrows more materials than those who come in the afternoon" by applying a DM algorithm.

Such data and knowledge acquired by analyzing the databases will become good tips for libraries to improve their patron services.

Figure 7 illustrates the cyclic structure of database and services. In the left part are databases in a library. Catalog database is constructed partly manually and partly automatically. Circulation database will be constructed automatically by using an AIDC technology. Personal data of patrons will be obtained when the library issues patron cards.

In the right part of the figure are library services. The OPAC service uses the materials' catalog data. In order to provide the "My Library" service, the system will use some patron data, circulation data, and others. The catalog data and other information will be well used for the reference service. The big arrow from databases and services indicate such used-use relationship.

On the other hand, we can collect some log data of services. From the OPAC service, we are able to have the data of which keywords were used, which library materials were chosen for getting detail data, and so on. From the My Library service, who login this service, when they use, how much time they use, which specific information they access, and so on.

These data themselves form another database, which is indicated the "Service Log" in Figure 7. We can use this database also for improving various services.

Take, for example, the OPAC service. Suppose a patron is trying to find appropriate keywords. When he or she types a keyword, it may be more convenient for the patron if the OPAC system shows some keywords relating to the given one. This is a

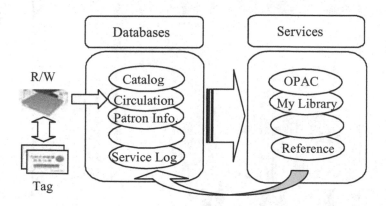

Fig. 7. Cyclic Structure of Database and Services

keyword recommendation [6] function. The keyword log data from the OPAC system itself provide its basis data. The system calculates the distance between two keywords in advance from this data. Then the recommended keywords are chosen among the keywords that are close to the keyword given by the patron.

If the My Library service provides a function of attaching comments to materials,this data can be used for OPAC service as well. Suppose we give a keyword and get a list of related library materials. The system can put a function of displaying what comments are attached on each material. The comments will help the patron with deciding if he or she reads it or not, reads in the library or borrows it when he/she decides to read

3.3 Improvement

As was mentioned in previous sections, RFID technology is very suitable to in-the-library marketing. By collecting and analyzing the data we can acquire useful knowledge that helps with providing better services to patrons, managing the library more efficiently and more effectively.

One important issue here is how to feedback the acquired knowledge to the patrons. An idea for improving the patron interface is the "virtual bookshelf [9, 13]." Figure 8 is an example screen. This is supposed to be a result of an OPAC system, in which the list of books is displayed with images of book spines. This way of display is much easier to intuitively recognize what books are in the list than that of display in text form. This is because we recognize and memorize books not only with title and/or authors, but also with the design of the book; color, font, arrangement, and so on.

One method of attaching extra information to each of the book is to put a link to the image of the spine of the book. If we click on a book spine, the information relating to the book will be displayed in a popup window. The information includes catalog data such as title, author, year of issue, etc. together with extra information such as comments by patrons and/or librarians.

Fig. 8. Virtual Bookshelf

Virtual bookshelf is also well used for personalized library services. Patrons may keep some number of his or her own virtual bookshelves in the library Web site. They can arrange the virtual books on whichever bookshelves they want. They are also able to rear-range the books whenever they want as if the bookshelves are their own and they are always at their study room. They can also put comments to the books from their personal views. The library system will collect such data and feedback the statistic results and other data including the evaluation data and comments to the patrons. One possible service is to recommend one or some books to the patrons that might be useful for the patrons to put in the patron's bookshelf.

4 Security Issue

Security issue is also important for RFID tag system [15]. In this section we discuss two types of security issues; how to protect library materials from theft and how to protect patron-related private data from leakage.

Take the first one. RFID tag system is used as a replacement of magnetic tag system, which is mostly used in university libraries in Japan. One reason of introducing the RFID tag system is to replace the magnetic tag for security and barcode book identifica-tion code with one RFID tag. This reduction of book processing cost somehow contrib-utes the reduction of the cost of installing the RFID tag systems and the cost of tags. This aspect belongs to the step 1 of library automation with RFID tag system.

For the aspect of the step2, RFID tag system has an advantage of having much detail data about materials and patrons. By utilizing this advantage the system analyzes the data and extracts information such as what types of materials are stolen than other types, what types of attributes correlate to possibility of being theft, and so on.

Take the second issue; i.e. protection from leakage of private data. Figure 9 illustrates one possible solution to this issue. The basic idea of this system is to put a separated pri-vate data management server (PDMS) and control the flow of data to and from the server.

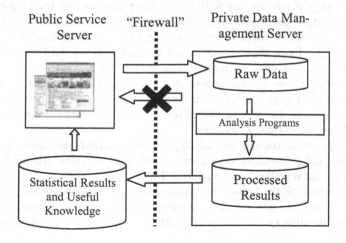

Fig. 9. Private Data Management System

The data of the library database that are needed in analyzing are stored in the IDMS. Only the manager operator of this machine can access to this original data. The analysis programs are invoked either automatically invoked programs or by an ordinary operator of this system. An ordinary operator is able to use only designated programs. The processed results, that are obtained by such analysis programs are accessible by the ordinary operators and thus they are allowed to be copied to outside of the machine. However, ordinary operators are not allowed to access to the original data in order to protect these data from leakage.

This mechanism is conceptually similar to firewall system for network traffic data. The major purpose of out mechanism is to protect data from leakage by operational errors.

5 Concluding Remarks

The background situation of this research lies that materials that are dealt with libraries are changing from physical ones such as books made of paper to digital ones. A big difference between these two materials is that digital materials are easy to deliver via network.

So the patrons do not need to visit a library to get materials. However, considering that we have a huge collection of books already, the library materials will be a mixed up of these two. So we expect the libraries last as hybrid libraries for a long time.

In this paper, we propose a model for hybrid libraries to transit seamlessly from the library where most materials are made of paper, i.e. books and journals, to the electric library, where most materials are provided as digital data. The key idea of this model is the good use of RFID tag system.

RFID is a representative technology of automatic identification and data capture (AIDC) technologies. With this technology, we have advantages of faster and multiple ID recognition, easy to use operational interface, etc., comparing to the barcode system, which is the currently dominating AIDC technology.

In the first step of introducing RFID for library automation, the major aim is to get efficiency and cost cutting. For example in a library of University of Connecticut, US [14], the number of counters can be reduced to half after installing RFID tag system. Even though the initial cost needed for RFID equipments is very high, the running cost is cheaper than that of barcode. However we would like to put strong stress on the importance of the second step for RFID in this paper. In the second step, we get extra data by utilizing RFID tag system and use the data for acquiring knowledge that is useful for improving library services (i.e. one of the library marketing [7] method).

The key equipment for the second step is RFID readers attached to the bookshelves, book trucks, browsing tables, and so on. By collecting and analyzing such data, we can extract statistical information and useful tips.

In order to feedback the knowledge to the user, we recommend the use of virtual shelf system. It is more intuitive and easy to recognize the books in a list. We have also illustrated that this system is a good platform to provide extra information to the patrons.

Lastly, we propose an system organization of private data management system. By using this model the raw data will be protected from leakage to public.

Currently the major reason for introducing RFID tag system is for efficiency. We are expecting and hoping that the major reason might change to for effectiveness in the near future.

Acknowledgments

I greatly acknowledge my co-researchers Prof Daisuke Ikeda of Kyushu University, Prof Takuya Kida of Hokkaido University, and Prof Kiyotaka Fujisaki for their discussions on digital libraries and RFID technologies. This research was partially supported by the Ministry of Education, Science, Sports and Culture, Grant-in-Aid for Scientific Research (B), 16300078, 2005.

References

1. AIREF Library: http://www.kenkou-fukuoka.or.jp/airef/tosyokan3.htm (in Japanese)
2. Baulkham Hills Shire Library Services: http://www.baulkhamhills.nsw.gov.au/library/
3. Finkenzeller, K.: RFID Handbook (Second Edition). John Wiley & Sons (2003)
4. Kyushu University Library: http://www.lib.kyushu-u.ac.jp/
5. Lee, Eung-Bong: Digital Library & Ubiquitous Library. Science and Technology Information Management Association Academic Seminar (V) (2004) (in Korean)
6. Oda, Mitsuru, Minami, Toshiro: From Information Search towards Knowledge and Skill Acquisition with SASS, Pacific Rim Knowledge Acquisition Workshop (PKAW2000), (2000)
7. Ohio Library Council: http://www.olc.org/marketing/
8. Minami, Toshiro: Needs and Benefits of Massively Multi Book Agent Systems for u-Libraries, T. Ishida, L. Gasser, and H. Nakashima(eds.), MMAS 2004, LNAI3446, Springer, (2005) 239-253
9. Minami, Toshiro: On-the-site Library Marketing for Patron Oriented Services. Bulletin of Kyushu Institute of Information Sciences, Vol.8 No.1 (2006) (in Japanese)
10. Ranganathan, S.R.: The Five Laws of Library Science, Bombay Asia Publishing House (1963)
11. Ramparany, F., Boissier, O.: Smart Devices Embedding Multi-Agent Technologies for a Pro-active World, Proc. Uniquitous Computing Workshop (2002)
12. Resnick, P. and Varian, H. R. (Guest Eds.): Recommender Systems. Communications of ACM, Vol. 40 No. 3 (1997) 56-89
13. Sugimoto, Shigeo et.al: Enhancing usability of network-based library information system - experimental studies of a user interface for OPAC and of a collaboration tool for library services. Proceedings of Digital Libraries '95 (1995) .115-122
14. University of Connecticut Libraries: http://spirit.lib.uconn.edu/
15. Weis, S.A., Sarma, S.E., Rivest, R.L., Engels, D.W.: Security and Privacy Aspects of Low-Cost Radio Frequency Identification Systems. Proc. First International Conference on Security in Pervasive Computing. Lecture Notes in Computer Science, Vol. 2802. Springer-Verlag (2003) 201-212

Extracting Discriminative Patterns from Graph Structured Data Using Constrained Search

Kiyoto Takabayashi[1], Phu Chien Nguyen[1], Kouzou Ohara[1], Hiroshi Motoda[2], and Takashi Washio[1]

[1] I.S.I.R., Osaka University,
8-1, Mihogaoka, Ibaraki, Osaka, 567-0047, Japan
{kiyoto_ra, chien, ohara, washio}@ar.sanken.osaka-u.ac.jp
[2] AFOSR/AOARD
7-23-17, Roppongi, Minato-ku, Tokyo 106-0032, Japan
hiroshi.motoda@aoard.af.mil

Abstract. A graph mining method, Chunkingless Graph-Based Induction (Cl-GBI), finds typical patterns appearing in graph-structured data by the operation called chunkingless pairwise expansion, or pseudo-chunking which generates pseudo-nodes from selected pairs of nodes in the data. Cl-GBI enables to extract overlapping subgraphs, but it requires more time and space complexities than the older version GBI that employs real chunking. Thus, it happens that Cl-GBI cannot extract patterns that need be large enough to describe characteristics of data within a limited time and given computational resources. In such a case, extracted patterns maynot be so interesting for domain experts. To mine more discriminative patterns which cannot be extracted by the current Cl-GBI, we introduce a search algorithm in which patterns to be searched are guided by domain knowledge or interests of domain experts. We further experimentally show that the proposed method can efficiently extract more discriminative patterns using a real world dataset.

1 Introduction

Over the last decade, there has been much research work on data mining which intend to find useful and interesting knowledge from massive data that are electronically available. A number of studies have been made in recent years especially on mining frequent patterns from graph-structured data, or simply graph mining, because of the high expressive power of graph representation [1,13,6,12,4,5].

Chunkingless Graph Based Induction (Cl-GBI) [8] is one of the latest algorithms in graph mining and an extension of Graph Based Induction (GBI) [13] that can extract typical patterns from graph-structured data by stepwise pair expansion, i.e., by recursively chunking two adjoining nodes. Similarly to GBI and its another extension, Beam-wise GBI(B-GBI) [6], Cl-GBI adopts the stepwise pair expansion principle, but never chunks adjoining nodes and contracts the graph. Instead, Cl-GBI regards a pair of adjoining nodes as a *pseudo-node*

A. Hoffmann et al. (Eds.): PKAW 2006, LNAI 4303, pp. 64–74, 2006.

and assigns a new label to it. This operation is called *pseudo-chunking* and can fully solve the reported problems caused by chunking, i.e., ambiguity in selecting nodes to chunk and incompleteness of the search. This is because every node is available to make a new pseudo-node at any time in Cl-GBI. However Cl-GBI requires more time and space complexities in exchange of gaining the ability of extracting overlapping patterns. Thus, it happens that Cl-GBI cannot extract patterns that need be large enough to describe characteristics of data within a limited time and a given computational resource. In such a case, extracted patterns may not be so much of interest for domain experts.

To improve the search efficiency, in this paper, we propose a method of guiding the search of Cl-GBI using domain knowledge or interests of domain experts. The basic idea is adopting patterns representing knowledge or interests of domain experts as constraints on the search, in order to effectively restrict the search space and extract more discriminative or interesting patterns than those which can be extracted by the current Cl-GBI. For that purpose, we use two types of patterns as the constraints: one is the pattern that should be included in the extracted ones, and the other is the pattern that should not be included in them. These patterns allow us to specify patterns of interest and patterns trivial for domain experts, respectively. We also experimentally show the effectiveness of the proposed search method by applying the constrained Cl-GBI to the hepatitis dataset which is a real world dataset.

In this paper, we deal with only connected labeled graphs, and use information gain [10] as the discriminativity criterion. In what follows, "a pair" denotes a pair of adjoining nodes in a graph.

2 Chunkingless Graph-Based Induction(Cl-GBI)

Stepwise pair expansion is an essential operation in GBI and its variants, which recursively generates new nodes from pairs of two adjoining nodes and links between them. In GBI, a pair is selected according to a certain criterion based on frequency, and all of its occurrences in graphs are replaced with a node having a newly assigned label. Namely each graph is rewritten each time a pair is chunked, and never restored in any subsequent chunking[1]. On one hand, this chunking mechanism is suitable for extraction of patterns from either a very large single graph or a set of graphs because extracted patterns can rapidly grow. On the other hand, it involves ambiguity in selecting nodes to chunk, which causes a crucial problem, i.e., possibility of overlooking some overlapping subgraphs due to inappropriate chunking order. Beam search adopted by B-GBI can alleviate this problem by chunking the b (beam width) most frequent pairs and copying each graph into respective states, but not completely solve it because chunking process is still involved.

---

[1] Note that this does not mean that the link information of the original graphs is lost. It is always possible to restore how each node is connected in the extracted subgraphs.

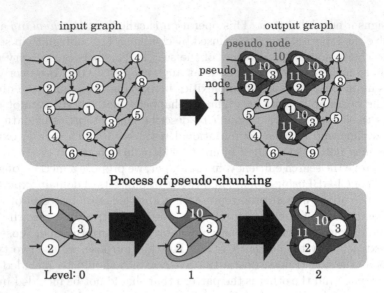

input graph

output graph

Process of pseudo-chunking

Level: 0   1   2

Fig. 1. An Example of Pseudo-chunking in Cl-GBI

In contrast to GBI and B-GBI, Cl-GBI does not chunk a selected pair, but regards it as a *pseudo-node* and assigns a new label to it. Thus, graphs are not "compressed" nor copied into respective states. Figure 1 illustrates examples of *pseudo-chunking* in Cl-GBI, in which a typical pattern consisting of nodes 1, 2, and 3 is extracted from the input graph. Cl-GBI first finds the pair $1 \to 3$ based on its frequency, and pseudo-chunks it, i.e., registers it as the pseudo-node 10, but does not rewrite the graph. Then, in the next iteration, it finds the pair $2 \to 10$, and pseudo-chunks and registers it as the pseudo-node 11. As a result, the typical pattern is extracted. In the rest of the paper, we refer to each iteration in Cl-GBI as "level".

The algorithm of Cl-GBI is shown in Fig.2. The search of Cl-GBI is controlled by the following parameters: a beam width b, the maximal number of levels of pseudo-chunking N, and a frequency threshold θ. In other words, at each level, the b most frequent pairs are selected from a set of pairs whose frequencies are not less than θ, and are pseudo-chunked.

3   Constrained Search for Cl-GBI

3.1   Patterns Used as Constraints

The current Cl-GBI blindly extracts a huge number of frequent pairs without any clues other than frequency and selects pairs to pseudo-chunk from among them. However, if the goal is finding patterns which are either discriminative or of interest for domain experts, the current method of extracting pairs is too naive and inefficient in both time and space complexities. This is because such

Input. A graph database D, a beam width b, the maximal number of levels of pseudo-chunking N, a frequency threshold θ

Output. A set of typical patterns S

Step 1. Extract all the pairs consisting of two connected nodes in the graphs, register their positions using node id (identifier) sets. From the 2nd level on, extract all the pairs consisting of two connected nodes with at least one node being a new pseudo-node.

Step 2. Count frequencies of extracted pairs and eliminate pairs whose frequencies count below θ.

Step 3. Select the b most frequent pairs from among the remaining pairs at Step 2 (from the 2nd level on, from among the unselected pairs in the previous levels and the newly extracted pairs). Each of the b selected pairs is registered as a new node. If either or both nodes of the selected pair are not original but pseudo-nodes, they are restored to the original patterns before registration.

Step 4. Assign a new label to each pair selected at Step 3 but do not rewrite the graphs. Go back to Step 1.

Fig. 2. Algorithm of Cl-GBI

patterns are not always frequent in a database. If discriminative patterns are not so frequent in a database and there are a large number of patterns that are more frequent than them, it is difficult to extract such discriminative patterns within a limited time and a given computational resource.

Note that such a goal, i.e., finding discriminative/interesting patterns is not special. In DT-ClGBI [9] which constructs a decision tree from graph-structured data, Cl-GBI is adopted to extract discriminative patterns used as test nodes in a tree. When we analyzed the hepatitis dataset [11] provided by Chiba University Hospital with Cl-GBI, domain experts (medical doctors) expected that patterns that are interesting for them were extracted, but in fact we could not find satisfactory ones with the current Cl-GBI.

Therefore, in this paper, we introduce domain knowledge or interests of domain experts and impose them as constraints on patterns that are extracted in Cl-GBI in order to efficiently extract discriminative/interesting patterns which the current Cl-GBI could not extract within a limited time and a given computational resource. We represent such domain knowledge and interests of domain experts as graphs, or patterns, and call them the *constraint patterns*.

Although various types of constraint patterns can be considered, in this paper, we focus on the following two types of patterns: one is the pattern that should be included in extracted patterns, and the other is the pattern that should not be included in them. We refer to the former type as the *INpattern*, and the latter as the *EXpattern*. Thus, the constraints we introduce in this paper are defined as follows:

Constraint 1. Extracted patterns must include INpatterns.

Constraint 2. Extracted patterns must not include EXpatterns.

Figures 3 (a) and (b) show the examples of Constraints 1 and 2, respectively. Note that in case of Constraint 1, not only patterns including the given INpat-

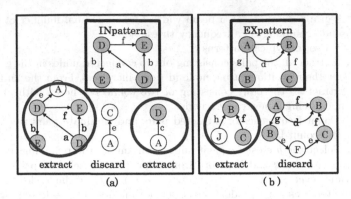

Fig. 3. Examples of INpatterns and EXpatterns

terns, but also patterns including at least one proper subgraph of the INpatterns should be extracted in order not to prevent patterns satisfying the imposed constraints from being generated in the succeeding steps based on the stepwise pair expansion principle. The case is illustrated at the bottom right in Fig.3 (a). We call such a pattern including only proper subgraphs of the given INpatterns a *neighborhood pattern*. In other words, an INpattern does not necessarily have to be identical to a pattern representing domain knowledge or interests of domain experts, but need merely include at least one of its proper subgraphs. Namely, Constraint 1 is useful to specify patterns of interest for domain experts and to aggressively search around them, while Constraint 2 is useful to avoid extracting trivial or boring patterns for domain experts.

In addition, in case of Constraint 1, we can discard a pair if it does not include any node/link labels appearing in the given INpatterns as shown in Fig.3 (a). This is because such a pair, or pattern can never grow to a pattern that includes at least one of the given INpatterns. Even if in fact the discarded pattern is a subgraph of a pattern P satisfying all constraints and a given evaluation criterion such as the minimal frequency, P can be constructed from another pattern including at least one node/link label appearing in the given INpatterns. In other words, enumerating only the pairs with at least one node/link label in the given INpatterns as candidates to be pseudo-chunked allows us to effectively restrict the search space. However, it is noted that a label with high frequency does not effectively work as a constraint because it may appear also in many pairs which the user intends to exclude from the search space. Thus, if a set of node (link) labels contains such frequent ones, we do not use the set to restrict pairs in the enumeration process.

3.2   Design of Constrained Search

To guide the search process of Cl-GBI using either INpatterns or EXpatterns, we have to check if an enumerated pair includes them, which requires additional subgraph isomorphism checking. Since subgraph isomorphism checking is known to be NP-complete [2], it is obvious that a naive approach could be computationally expensive. To reduce the computational cost as much as possible, we should detect pairs that have no possibility of including constraint patterns before the checking. For that purpose, we focus on the quantities that characterize the topological structure of a graph, especially the number of node/link labels in patterns.

First of all, for two arbitrary pairs, or patterns x and y, we define a quantity T_{num} as follows:

$$T_{num}(x,y) = \sum_{L_k \in L(y)} f(x, L_k), \tag{1}$$

where $L(y)$ is a set of labels appearing in y, and $f(x, L_k)$ is the number of occurrences of the label $L_k \in L(y)$ in x. Note that if x is identical to y, $T_{num}(x, y)$ must be equal to $T_{num}(y, y)$. Similarly, $T_{num}(x, y)$ must be greater than $T_{num}(y, y)$ if y is a subgraph of x. Consequently, we can skip subgraph isomorphism checking for the pair of an enumerated pattern P_i and a constraint pattern T_j if $T_{num}(P_i, T_j) < T_{num}(T_j, T_j)$ because P_i never includes T_j.

Furthermore, we can prune more subgraph isomorphism checking even if $T_{num}(P_i, T_j) \geq T_{num}(T_j, T_j)$ because it does not guarantee that P_i necessarily includes T_j. In order for P_i to include T_j, for every label appearing in T_j, the number of its occurrences in P_i has to be greater than or equal to that in T_j. Namely, we can skip subgraph isomorphism checking for P_i if P_i does not satisfy this condition. To check this condition, we define the following boolean value P_{info} for two patterns x and y.

$$P_{info}(x, y) = \bigwedge_{L_k \in L(y)} p(x, y, L_k), \tag{2}$$

where

$$p(x, y, L_k) = \begin{cases} true & \text{if } f(x, L_k) \geq f(y, L_k), \\ false & \text{otherwise.} \end{cases}$$

If $P_{info}(P_i, T_j)$ is $true$, then subgraph isomorphism checking has to be done; otherwise it can be skipped.

These ideas discussed above are summarized in Fig.4 as the algorithm that extracts pairs, which is invoked at **Step 1** in the algorithm shown in Fig.2 and provides a set of candidate pairs to be pseudo-chunked at each level. As input, a graph database D, a set of constraint patterns T consisting of either INpatterns or EXpatterns, a parameter L_v specifying the current level, and a list of pairs L consisting of patterns that have been extracted before are given. Then it outputs L adding newly extracted pairs. $PD(P_i, T_j)$ in this algorithm is the procedure for subgraph isomorphism checking, which returns true if P_i includes T_j; otherwise false. In fact, it applies Cl-GBI to a graph database consisting of only P_i and T_j after deleting all nodes and links that never appear in T_j from P_i. By running Cl-GBI without the constraint patterns until T_j is extracted and checking the occurrences of T_j in P_i, one can detect whether P_i includes T_j or not.

ExtPair(D, T, L, L_v, M)
Input: a database D, a set of constraint patterns T, the current level L_v,
        a set of extracted pairs L (initially empty),
        the constraint mode M (either "INpattern" or "EXpattern");
Output: a set of extracted pairs L with newly extracted pairs;
begin
  if $L_v = 1$ then
    if $M =$ "INpattern" then
      Enumerate pairs in D, which consist of nodes or links
      appearing in T, and store them in E;
    else
      Enumerate all the pairs in D and store them in E;
  else
    Enumerate pairs, which consist of one or both
    pseudo-nodes in L, and store them in E;
  for each $P_i \in E$ begin
    if P_i is marked then
      $L := L \cup \{P_i\}$; next;
    register := 1;
    for each $T_j \in T$ begin
      if $T\_num(P_i, T_j) \geq T\_num(T_j, T_j)$ then
        if $P_{info}(P_i, T_j) =$ **true** then
          if $M =$ "INpattern" then
            if $PD(P_i, T_j) =$ **true** then mark P_i;
          else
            if $PD(P_i, T_j) =$ **true** then
              discard P_i; *register* := 0; **break**;
    end
    if *register* = 1 then $L := L \cup \{P_i\}$;
  end
  return L;
end

Fig. 4. Algorithm of the constrained pattern extraction

4   Experimental Evaluation

4.1   Experimental Settings

To evaluate the proposed method, we implemented Cl-GBI with the algorithm
shown in Fig.4 on PC (CPU: Pentium 4 3.2GHz, Memory: 4GB, OS: Fedora
Core release 3) in C++, and applied this *constrained Cl-GBI* to the hepatitis
dataset. The current system has a limitation that either INpattern constraints
or EXpattern constraints can be imposed at a time. In this experiment, we used
two classes, *Response* and *Non-Response*, in the dataset, denoted by R and N,
respectively. R consists of patients to whom the interferon therapy was effective,
while N consists of those to whom it was not effective. We used 24 inspection

Table 1. Size of graphs of the hepatitis dataset

class	R	N
number of graphs	38	56
average number of nodes in a graph	104	112
maximal number of nodes in a graph	145	145
minimum number of nodes in a graph	24	20
total number of nodes	3,944	6,296
kinds of node labels	12	
average number of links in a graph	108	117
maximal number of links in a graph	154	154
minimum number of links in a graph	23	19
total number of links	4,090	6,577
kinds of link labels	30	

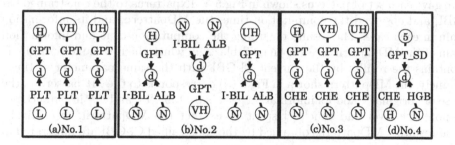

Fig. 5. INpatterns used in the experiments

items as attributes, and converted the records of each patient into a graph in the same way as [3]. The statistics on the size of resulting graphs are shown in Table. 1.

In this experiment, we used 4 sets of INpatterns shown in Fig.5, in which (a) to (c) are patterns reported in [7] and represent typical examination results for patients belonging to R, while (d) is the pattern with the highest information gain, or the most discriminative pattern among ones extracted by the current Cl-GBI. In the following, we refer to the pattern which is the most discriminative one among extracted patterns as the *MDpattern*. The node with the label "d" in Fig.5 is a dummy node representing a certain point of time. For example, the leftmost pattern in Fig.5 (a) means that at a certain point of time, the value of GPT (glutamic-pyruvic transaminase) is High and the value of PLT (platelet) is Low. Note that node labels in this dataset such as "d", "H", etc. are common and may appear with large frequency. Thus, we used only the link labels appearing in the INpatterns as constraints for the pair enumeration as discussed above. As for the parameters of Cl-GBI, we set them as follows: $b = 10$, $N = 10$, and $\theta = 0\%$.

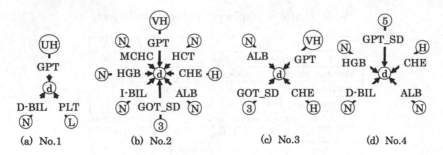

Fig. 6. MDpatterns Extracted by the constrained Cl-GBI

4.2 Experimental Results

We gave each set of patterns shown in Fig.5 as INpatterns to the constrained Cl-GBI, and observed the computation time, the MDpattern, and its information gain in each case. The results regarding the computation time and information gain of the MDpatterns are shown in Table. 2, in which the row named "original" contains the results by the current Cl-GBI with the same parameter settings. Namely, the MDpattern shown in Fig.5 (d) corresponds to the MDpattern in the case of "original", and its information gain is 0.1139. "L" and "t" in parentheses denote the level and time[sec] spent to extract the MDpattern, respectively. The resulting MDpatterns obtained by the constrained Cl-GBI are illustrated in Fig.6.

First, focusing on the column of "max information gain" in Table 2, it is found that the MDpatterns extracted by the constrained Cl-GBI are more discriminative than the MDpattern by the current Cl-GBI in the cases of No.2 and No.4. In the cases of No.1 and No.3, the values of information gain of the MDpatterns are compatible with that of the MDpattern extracted by the current Cl-GBI. In addition, the computation times in all 4 cases using INpatterns are much less than in the case of the current Cl-GBI. Note that the values in parentheses in the column "time" of Table 2 are corresponding computation times by the constrained Cl-GBI without pruning subgraph isomorphism checking. Comparing with computation times in the same row, it could be said that checking if pairs include constraint patterns based on the number of nodes/links could

Table 2. Experimental results

	time[sec]	max information gain
original	44,973 (—)	0.1139 (L:4, t:1292)
No.1	9,355 (66,211)	0.1076 (L:3, t:18)
No.2	6,893 (31,527)	0.1698 (L:5, t:376)
No.3	20,495 (159,434)	0.1110 (L:3, t:55)
No.4	4,970 (14,923)	0.1297 (L:4, t:39)

reduce the computation time significantly. From these results, we can say that given appropriate constraints, the constrained Cl-GBI could efficiently extract patterns which are more discriminative than those which are extracted by the current Cl-GBI.

Next, as shown in Fig.6, except in the case of No.3, the MDpatterns extracted by the constrained Cl-GBI include one of the given INpatterns completely. In the case of No.3, the MDpattern includes a subgraph of one of the INpatterns. Thus, it is expected that the proposed constrained search method may work well even if it is not sure that given INpatterns are genuinely appropriate ones. This is because it can extract not only patterns completely including them, but also the neighborhood patterns.

In addition, note that the INpattern used in the case of No.4 is the MDpattern obtained by the current Cl-GBI with the same parameter settings, and works as a good constraint, succeeding in extracting more discriminative patterns. From this result, it is expected that MDpatterns obtained before may work as good constraints to guide the search, which would be desirable if no domain knowledge is available to restrict the search space: in such a case, instead of running the current Cl-GBI only once setting the maximal level L to a large value, repeatedly running the constrained Cl-GBI setting L to a smaller value and using the MDpattern extracted by the previous run as the new INpattern might allow us to extract patterns that are more discriminative in a less computation time. Verifying this expectation is one of our future work.

5   Conclusion

In this paper, we proposed a constrained search method that effectively restricts the search space of Cl-GBI by imposing domain knowledge or interests of domain experts as constraints on patterns to be searched, and embedded it in Cl-GBI, resulting in the constrained Cl-GBI. The proposed method avoids conducting subgraph isomorphism checking as much as possible based on the number of node/link labels in patterns because it is computationally expensive. Experimental results showed that if given constraints are appropriate, the constrained Cl-GBI can extract more discriminative patterns in a less computation time than the current Cl-GBI. In addition, the results also showed the possibility that discriminative patterns extracted in earlier steps in the search may work as good constraints in the constrained Cl-GBI and contribute to extracting more discriminative patterns. It is worth saying that the basic idea of imposing constraints on extracting patterns in graph mining does not rely on how to construct candidate patterns. Namely, this approach could apply to any grpah mining method if the patterns were constructed step by step.

As future work, we plan to provide more flexible ways to give constraints such as a combination of INpatterns and EXpatterns, and to further evaluate the constrained Cl-GBI by comparing it with other graph mining methods including ones based on Inductive Logic Programming. In addition, we need to verify the resulting patterns coorporating with domain experts such as medical doctors.

References

1. Cook, D. J. and Holder, L. B.: Substructure Discovery Using Minimum Description Length and Background Knowledge. Artificial Intelligence Research, Vol. 1, pp. 231–255, (1994).
2. Fortin, S.: The Graph Isomorphism Problem. Technical Report TR96-20, Department of Computer Science, University of Alberta, (1996).
3. Geamsakul, W., Yoshida, T., Ohara, K., Motoda, H., Yokoi, H., and Takabayashi, K.: Constructing a Decision Tree for Graph-Structured Data and its Applications. Fundamenta Informaticae Vol. 66, No.1-2, pp. 131–160, (2005).
4. Inokuchi, A., Washio, T., and Motoda, H.: Complete Mining of Frequent Patterns from Graphs: Mining Graph Data. Machine Learning, Vol. 50, No. 3, pp. 321–354, (2003).
5. Kuramochi, M. and Karypis, G.: An Efficient Algorithm for Discovering Frequent Subgraphs. IEEE Trans. Knowledge and Data Engineering, Vol. 16, No. 9, pp. 1038–1051, (2004).
6. Matsuda, T., Motoda, H., Yoshida, T., and Washio, T.: Mining Patterns from Structured Data by Beam-wise Graph-Based Induction. Proc. of DS 2002, pp. 422–429, (2002).
7. Motoyama, S., Ichise, R., and Numao, M.: Knowledge Discovery from Inconstant Time Series Data. JSAI Technical Report, SIG-KBS-A405, pp. 27–32, in Japanese, (2005).
8. Nguyen, P. C., Ohara, K., Motoda, H., and Washio, T.: Cl-GBI: A Novel Approach for Extracting Typical Patterns from Graph-Structured Data. Proc. of PAKDD 2005, pp. 639–649, (2005).
9. Nguyen, P. C., Ohara, K., Mogi, A., Motoda, H., and Washio, T.: Constructing Decision Trees for Graph-Structured Data by Chunkingless Graph-Based Induction. Proc. of PAKDD 2006, pp. 390–399, (2006).
10. Quinlan, J. R.: Induction of decision trees. Machine Learning, Vol. 1, pp. 81–106, (1986).
11. Sato, Y., Hatazawa, M., Ohsaki, M., Yokoi, H., and Yamaguchi, T.: A Rule Discovery Support System in Chronic Hepatitis Datasets. First International Conference on Global Research and Education (Inter Academia 2002), pp. 140–143, (2002).
12. Yan, X. and Han, J.: gSpan: Graph-Based Structure Pattern Mining. Proc. of the 2nd IEEE International Conference on Data Mining (ICDM 2002), pp. 721–724, (2002).
13. Yoshida, K. and Motoda, H.: CLIP: Concept Learning from Inference Patterns. Artificial Intelligence, Vol. 75, No. 1, pp. 63–92, (1995).

Evaluating Learning Algorithms with Meta-learning Schemes for a Rule Evaluation Support Method Based on Objective Indices

Hidenao Abe[1], Shusaku Tsumoto[1], Miho Ohsaki[2],
Hideto Yokoi[3], and Takahira Yamaguchi[4]

[1] Department of Medical Informatics, Shimane University, School of Medicine
89-1 Enya-cho, Izumo, Shimane 693-8501, Japan
abe@med.shimane-u.ac.jp, tsumoto@computer.org
[2] Faculty of Engineering, Doshisha University
mohsaki@mail.doshisha.ac.jp
[3] Department of Medical Informatics, Kagawa University Hospital
yokoi@med.kagawa-u.ac.jp
[4] Faculty of Science and Technology, Keio University
yamaguti@ae.keio.ac.jp

Abstract. In this paper, we present evaluations of learning algorithms for a novel rule evaluation support method in data mining post-processing, which is one of the key processes in a data mining process. It is difficult for human experts to evaluate many thousands of rules from a large dataset with noises completely. To reduce the costs of rule evaluation task, we have developed the rule evaluation support method with rule evaluation models, which are learned from a dataset consisted of objective indices and evaluations of a human expert for each rule. To enhance adaptability of rule evaluation models, we introduced a constructive meta-learning system to choose proper learning algorithms for constructing them. Then, we have done a case study on the meningitis data mining result, the hepatitis data mining results and rule sets from the eight UCI datasets.

1 Introduction

In recent years, with huge data stored on information systems in natural science, social science and business domains, developing information technologies, people hope to find out valuable knowledge suited for their purposes. Besides, data mining techniques have been widely known as a process for utilizing stored data on database systems, combining different kinds of technologies such as database technologies, statistical methods and machine learning methods. In particular, if-then rules are discussed as one of highly usable and readable output of data mining. However, to large dataset with hundreds attributes including noise, the process often obtains many thousands of rules, which rarely include valuable rules for a human expert.

A. Hoffmann et al. (Eds.): PKAW 2006, LNAI 4303, pp. 75–88, 2006.

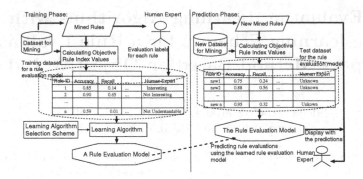

Fig. 1. Overview of the construction method of rule evaluation models

To support such a rule selection, many efforts have done using objective rule evaluation indices such as recall, precision, and other interestingness measurements [16,30,33] (we call them "objective indices" later). However, it is also difficult to estimate a criterion of a human expert with single objective rule evaluation index, because his/her subjective criterion such as interestingness and importance for his/her purpose is influenced by the amount of his/her knowledge and/or a passage of time.

To above issues, we have been developed an adaptive rule evaluation support method for human experts with rule evaluation models, which predict experts' criteria based on objective indices, re-using results of evaluations of human experts. In Section 2, we describe the rule evaluation model construction method based on objective indices. Since our method needs more accurate rule evaluation model to support a human expert more exactly, we present a performance comparison of learning algorithms for constructing rule evaluation models in Section 3. With the results of the comparison, we present the availability of learning algorithms from a constructive meta-learning system[1] for our rule evaluation model construction approach.

2   Rule Evaluation Support with Rule Evaluation Model Based on Objective Indices

We considered the process of modeling rule evaluations of human experts as the process to clear up relationships between the human evaluations and features of input if-then rules. With this consideration, we decided that the process of rule evaluation model construction can be implemented as a learning task. Fig.1 shows the process of rule evaluation model construction based on re-use of human evaluations and objective indices for each mined rule.

At the training phase, attributes of a meta-level training data set is obtained by objective indices such as recall, precision and other rule evaluation values. The human evaluations for each rule are joined as class of each instance. To obtain this data set, a human expert has to evaluate the whole or part of input rules at least once. After obtaining the training data set, its rule evaluation model is constructed by a learning algorithm. At the prediction phase, a human expert receives predictions for new rules based on their values of the objective indices. Since the task of rule evaluation models is a prediction, we need to choose a learning algorithm with higher accuracy as same as current classification problems.

3   Performance Comparisons of Learning Algorithms for Rule Model Construction

To predict human evaluation labels of a new rule based on objective indices more exactly, we have to construct a rule evaluation model, which has higher predictive accuracy.

In this section, we firstly present the results of an empirical evaluation with the dataset from the result of a meningitis data mining [14], hepatitis data mining [22,2] and that of the eight rule sets from eight UCI benchmark datasets [15]. With the experimental results, we discuss about the following three view points: performances of rule evaluation models, minimum training subset to construct a valid rule evaluation model, and contents of learned rule evaluation models.

As evaluations of performances of rule evaluation models, we have compared predictive accuracies on the whole dataset and Leave-One-Out. The accuracy of a validation dataset D is calculated with correctly predicted instances $Correct(D)$ as $Acc(D) = (Correct(D)/|D|) \times 100$, where $|D|$ means the size of the dataset. Recalls of class i on a validation dataset is calculated with correctly predicted instances about the class $Correct(D_i)$ as $Recall(D_i) = (Correct(D_i)/|D_i|) \times 100$, where $|D_i|$ means the size of instances with class i. Also the precision of class i is calculated with the size of instances predicted i as $Precision(D_i) = (Correct(D_i)/Predicted(D_i)) \times 100$.

As for estimating minimum training subset to construct a valid rule evaluation model, we obtained learning curves about accuracies to the whole training dataset to evaluate whether each learning algorithm can perform in early stage of a process of rule evaluations.

On the result of the actual data mining, we have investigated elements of the rule evaluation models. Then, we consider the characteristics of objective indices, which are used in these rule evaluation models.

To construct a dataset to learn a rule evaluation model, values of objective indices have been calculated for each rule, taking 39 objective indices as shown in Table1. Thus, each dataset for each rule set has the same number of instances as the rule set. Each instance consists of 40 attributes including the class attribute.

To these dataset, we applied nine learning algorithms to compare their performance as a rule evaluation model construction method. We have taken the following learning algorithms from Weka [31]: C4.5 decision tree learner [27] called J4.8, neural network learner with back propagation (BPNN) [17], support

Table 1. The objective rule evaluation indices for classification rules used in this research. **P:** Probability of the antecedent and/or consequent of a rule. **S:** Statistical variable based on P. **I:** Information of the antecedent and/or consequent of a rule. **N:** Number of instances included in the antecedent and/or consequent of a rule. **D:** Distance of a rule from the others based on rule attributes.

Theory	Index Name (**Abbreviation**) [Reference Number of Literature]
P	Coverage(**Coverage**), Prevalence(**Prevalence**)
	Precision(**Precision**), Recall(**Recall**)
	Support(**Support**), Specificity(**Specificity**)
	Accuracy(**Accuracy**), Lift(**Lift**)
	Leverage(**Leverage**), Added Value(**Added Value**)[30]
	Klösgen's Interestingness(**KI**)[19], Relative Risk(**RR**)[3]
	Brin's Interest(**BI**)[6], Brin's Conviction(**BC**)[6]
	Certainty Factor(**CF**)[30], Jaccard Coefficient(**Jaccard**)[30]
	F-Measure(**F-M**)[28], Odds Ratio(**OR**)[30]
	Yule's Q(**YuleQ**)[30], Yule's Y(**YuleY**)[30]
	Kappa(**Kappa**)[30], Collective Strength(**CST**)[30]
	Gray andOrlowska's Interestingness weighting Dependency(**GOI**)[12]
	Gini Gain(**Gini**)[30], Credibility(**Credibility**)[13]
S	χ^2 Measure for One Quadrant(χ^2-**M1**)[11]
	χ^2 Measure for Four Quadrant(χ^2-**M4**)[11]
I	J-Measure(**J-M**)[29], K-Measure(**K-M**)[23]
	Mutual Information(**MI**)[30]
	Yao and Liu's Interestingness 1 based on one-way support(**YLI1**)[33]
	Yao and Liu's Interestingness 2 based on two-way support(**YLI2**)[33]
	Yao and Zhong's Interestingness(**YZI**)[33]
N	Cosine Similarity(**CSI**)[30], Laplace Correction(**LC**)[30]
	ϕ Coefficient(ϕ)[30], Piatetsky-Shapiro's Interestingness(**PSI**)[24]
D	Gago and Bento's Interestingness(**GBI**)[10]
	Peculiarity(**Peculiarity**)[34]

vector machines (SVM)[1][25], classification via linear regressions (CLR)[2][7], and OneR [18]. In addition, we have also taken the following selective meta-learning algorithms: Bagging [5], Boosting [9] and Stacking[3] [32].

3.1　A Case Study on the Meningitis Datamining Result

In this case study, we have taken 244 rules, which are mined from six datasets about six kinds of diagnostic problems as shown in Table 2. These datasets are consisted of appearances of meningitis patients as attributes and diagnoses for each patient as class. Each rule set was mined with each proper rule induction algorithm composed by a constructive meta-learning system called CAMLET [14]. For each rule, we labeled three evaluations (I: Interesting, NI: Not-Interesting, NU: Not-Understandable), according to evaluation comments from a medical expert.

Constructing a Proper Learning Algorithm to Construct the Meningitis Rule Evaluation Model. We have developed a constructive meta-learning

---

[1] The kernel function was set up polynomial kernel.

[2] We set up the elimination of collinear attributes and the model selection with greedy search based on Akaike Information Metric.

[3] This stacking has taken the other seven learning algorithms as base-level learner and J4.8 as meta-level learner.

Table 2. Description of the meningitis datasets and their datamining results

Dataset	#Attributes	#Class	#Mined rules	#'I' rules	#'NI' rules	#'NU' rules
Diag	29	6	53	15	38	0
C_Cource	40	12	22	3	18	1
Culture+diag	31	12	57	7	48	2
Diag2	29	2	35	8	27	0
Course	40	2	53	12	38	3
Cult_find	29	2	24	3	18	3
TOTAL	—	—	244	48	187	9

system called CAMLET [1] to choose a proper learning algorithm to a given dataset with machine learning method repository. To implement the method repository, firstly, we identified each functional part called method from the following eight learning algorithms: Version Space [21], AQ15 [20], Classifier Systems [4], Neural Network, ID3 [26], C4.5, Bagging and Boosting. With the method repository CAMLET constructs a proper learning algorithm to a given dataset, searching possible learning algorithm specification space which is obtained by the method repository.

Since we have set up the number of refinement $N = 100$, CAMLET searched up to 400 learning algorithms from 6000 possible learning algorithms for the best one. Fig. 2 shows the constructed algorithm by CAMLET to the dataset of meningitis datamining result.

This algorithm iterates boosting of C4.5 decision tree for randomly split training datasets. Each classifier set generated by C4.5 decision tree learner is reinforced with the method from Classifier Systems. Then, the learned committee aggregates with weighted voting from boosting.

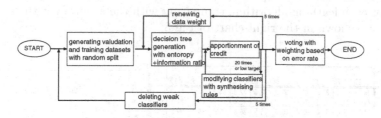

Fig. 2. The learning algorithm constructed by CAMLET for the dataset of the meningitis datamining result

Comparison on Classification Performances. In this section, we show the result of the comparisons of accuracies on the whole dataset, recall of each class label, and precisions of each class label.

The results of the performances of the nine learning algorithms to the whole training dataset and the results of Leave-One-Out are also shown in Table 3. All of the accuracies, Recalls of I and NI, and Precisions of I and NI on the whole training dataset are higher than just predicting each label at random.

Table 3. Accuracies (%), Recalls (%) and Precisions (%) of the nine learning algorithms

Learning Algorithms	Acc.	Evaluation on the training dataset					
		Recall			Precision		
		I	NI	NU	I	NI	NU
CAMLET	89.4	70.8	97.9	11.1	85.0	90.2	100.0
Stacking	81.1	37.5	96.3	0.0	72.0	87.0	0.0
Boosted J4.8	99.2	97.9	99.5	100.0	97.9	99.5	100.0
Bagged J4.8	87.3	62.5	97.9	0.0	81.1	88.4	0.0
J4.8	85.7	41.7	97.9	66.7	80.0	86.3	85.7
BPNN	86.9	81.3	89.8	55.6	65.0	94.9	71.4
SVM	81.6	35.4	97.3	0.0	68.0	83.5	0.0
CLR	82.8	41.7	97.3	0.0	71.4	84.3	0.0
OneR	82.0	56.3	92.5	0.0	57.4	87.8	0.0

Learning Algorithms	Acc.	Leave−One−Out(LOO)					
		Recall			Precision		
		I	NI	NU	I	NI	NU
CAMLET	80.3	7.4	73.0	0.0	7.4	73.0	0.0
Stacking	81.1	37.5	96.3	0.0	72.0	87.0	0.0
Boosted J4.8	74.2	37.5	87.2	0.0	39.1	84.0	0.0
Bagged J4.8	77.9	31.3	93.6	0.0	50.0	81.8	0.0
J4.8	79.1	29.2	95.7	0.0	63.6	82.5	0.0
BPNN	77.5	39.6	90.9	0.0	50.0	85.9	0.0
SVM	81.6	35.4	97.3	0.0	68.0	83.5	0.0
CLR	80.3	35.4	95.7	0.0	60.7	82.9	0.0
OneR	75.8	27.1	92.0	0.0	37.1	82.3	0.0

The accuracies of Leave-One-Out show robustness of each learning algorithm by which have been achieved from 75.8% to 81.9%.

The learning algorithm constructed by CAMLET shows the second accuracy to the whole training dataset, comparing with other learning algorithms. Thus, CAMLET shows higher adaptability than the other selective meta-learning algorithms.

Estimating Minimum Training Subsets for Each Learning Algorithms. The left table in Fig.3 shows accuracies to the whole training dataset with each subset of training dataset. Each data point is averaged accuracies from 10 times trials of randomly sub-sampled training datasets. The percentages of achievements for each learning algorithm, comparing with the accuracy with the whole dataset, are shown in the right chart of Fig.3.

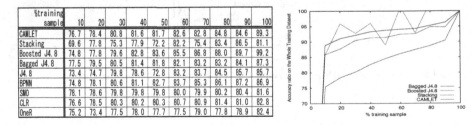

%training sample	10	20	30	40	50	60	70	80	90	100
CAMLET	76.7	78.4	80.8	81.6	81.7	82.6	82.8	84.8	84.6	89.3
Stacking	69.6	77.8	75.3	77.9	72.2	82.2	75.4	83.4	86.5	81.1
Boosted J4.8	74.8	77.8	79.6	82.8	83.6	85.5	86.8	88.0	89.7	99.2
Bagged J4.8	77.5	79.5	80.5	81.4	81.8	82.1	83.2	83.2	84.1	87.3
J4.8	73.4	74.7	79.8	78.6	72.8	83.2	83.7	84.5	85.7	85.7
BPNN	74.8	78.1	80.6	81.1	82.7	83.7	85.3	86.1	87.2	86.9
SMO	78.1	78.6	79.8	79.8	79.8	80.0	79.9	80.2	80.4	81.6
CLR	76.6	78.5	80.3	80.2	80.3	80.7	80.9	81.4	81.0	82.8
OneR	75.2	73.4	77.5	78.0	77.7	77.5	79.0	77.8	78.9	82.4

Fig. 3. Accuracies (%) with training sub-samples to the whole training dataset on the left table. And the chart of achieve rates(%) to the accuracies with the whole training dataset on the meta-learning algorithms.

As shown in these results, SVM, CLR and bagged J4.8 achieves higher than 95% with only less than 10% of training subset. Looking at the result of learning algorithm constructed by CAMLET, this algorithm achieves almost as same

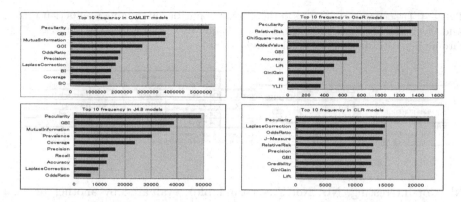

Fig. 4. Top 10 of frequencies of indices used in models of each learning algorithm with 10000 bootstrap samples of the meningitis datamining result dataset and executions

performance as bagged J4.8 with smaller training subset. However, it can outperform bagged J4.8 with larger training subsets. Although the constructed algorithm based on boosting, the combination of reinforcement method from Classifier Systems and the outer loop has been able to overcome a disadvantage of boosting for smaller training subset.

Rule Evaluation Models on the Meningitis Datamining Result Dataset. In this section, we present the statistics of rule evaluation models to the 10000 bootstrap re-sampled dataset learned with the algorithm constructed by CAMLET, OneR, J4.8 and CLR, because they are represented as explicit models such as a rule set, a decision tree, and a set of linear models.

As shown in Fig. 4, indices used in learned rule evaluation models are not only the group of indices increasing with a correctness of a rule, but also they are used some different groups of indices on different models. Almost indices such as YLI1, Laplace Correction, Accuracy, Precision, Recall, Coverage, PSI and Gini Gain are the former type of indices on the models. The later indices are GBI and Peculiality, which sums up difference of antecedents between one rule and the other rules in the same rule set.

3.2   A Case Study on the Chronic Hepatitis Datamining Results

In this case study, we have taken four datamining results about chronic hepatitis as shown in the left table of Table 4. These datasets are consisted of patterns for each laboratory test value about blood and urine of chronic hepatitis patients as attributes. Firstly, we have done datamining processes to find out relationships between patterns of attributes and patterns of GPT as class, which is one of the important test items to grasp conditions of each patient, two times. Second, we have also done other datamining processes to find out relationships between patterns of attributes and results of interferon (IFN) therapy two times. For each rule, we labeled three evaluations (EI: Especially Interesting, I: Interesting,

Table 4. Description of datasets of the chronic hepatitis datamining results (left table). And Overview of constructed learning algorithms by CAMLET to the datasets of the chronic hepatitis datamining results (right table).

	#Rules	Class Distribution				%Def class
		EI	I	NI	NU	
GPT						
Phase1(GPT1)	30	3	8	16	3	53.33
Phase2(GPT2)	21	2	6	12	1	57.14
IFN						
First Time(IFN1)	26	4	7	11	7	42.31
Second Time(IFN2)	32	15	5	11	1	46.88

	original classifier set	overall control structure	final eval. method
GPT1	C4.5 tree	Bagging	Best selection
GPT2	C4.5 tree	CS+Boost+Iteration	Weighted Voting
IFN1	C4.5 tree	CS+Boost+Iteration	Weighted Voting
IFN2	C4.5 tree	CS+Boost+Iteration	Weighted Voting

CS means including reinfoecement of classifier set from Classifiser Systems
Boost means including methods and control structure from Boosting

NI: Not-Interesting, NU: Not-Understandable), evaluated by another medical expert.

Constructing Proper Learning Algorithms for Chronic Hepatitis Datamining Results. As same as the construction of the proper learning algorithm for the meningitis data mining result, we constructed proper learning algorithms for the datasets of the four chronic hepatitis datamining results. The right table of Table 4 shows an overview of constructed learning algorithms for each dataset.

Comparison on Classification Performances. The results of the performances of the nine learning algorithms to the whole training dataset and the results of Leave-One-Out are shown in Table5. Almost of the accuracies on the whole training dataset are higher than just predicting each default class. The accuracies of Leave-One-Out show robustness of each learning algorithm. To GPT1 and IFN1, they are lower than just predicting default classes, because the medical expert evaluated these datamining results without certain criterion in his mind.

The learning algorithm constructed by CAMLET shows almost the same predictive performance as Boosted J4.8 on LOO, because these algorithms consist of C4.5 decision tree learner with boosting control structure.

Estimating Minimum Training Subset to Construct a Valid Rule Evaluation Model. As same as the case study of meningitis dataset, we have estimated the minimum training subsets for a valid model, which works better than just predicting a default class as shown in Table 5.

To GPT1 and IFN1, these algorithms need more instances to learn valid rule evaluation models than that of GPT2 and IFN2. This caused by the difference of human criteria when evaluating each datamining result.

Rule Evaluation Models on the Chronic Hepatitis Datamining Result Dataset. In this section, we present the statistics of rule evaluation models to the 10000 times bootstrap re-sampled dataset learned with the algorithm constructed by CAMLET to compare the difference among the models.

As shown in Fig. 5, these models consist of not only indices expressing correctness of rules but also other types of indices as shown in meningitis rule evaluation

Table 5. Accuracies(%), Recalls(%) and Precisions(%) of the nine learning algorithms on training dataset(the left table) and Leave-One-Out(the center table). Minimum training instances to construct valid rule evaluation models with each learning algorithm (the right table).

	On Training									Leave-One-Out									Min.
	Acc	Precision				Recall				Acc	Precision				Recall				Estim
		EI	I	NI	NU	EI	I	NI	NU		EI	I	NI	NU	EI	I	NI	NU	
GPT1																			
J4.8	96.7	100.0	88.9	100.0	100.0	66.7	100.0	100.0	100.0	50.0	0.0	60.0	60.0	0.0	0.0	75.0	56.3	0.0	14
BPNN	100.0	100.0	100.0	100.0	100.0	100.0	100.0	100.0	100.0	30.0	0.0	12.5	50.0	0.0	0.0	12.5	50.0	0.0	14
SVM	56.7	0.0	100.0	68.2	14.3	0.0	12.5	93.8	33.3	46.7	0.0	0.0	65.0	11.1	0.0	0.0	81.3	33.3	20
CLR	63.3	0.0	66.7	62.5	0.0	0.0	50.0	93.8	0.0	40.0	0.0	14.3	50.0	0.0	0.0	12.5	68.8	0.0	16
OneR	60.0	0.0	66.7	59.3	0.0	0.0	25.0	100.0	0.0	43.3	0.0	25.0	55.6	0.0	0.0	37.5	62.5	0.0	14
BagJ4.8	93.3	75.0	87.5	100.0	100.0	100.0	87.5	93.8	100.0	33.3	0.0	12.5	50.0	0.0	0.0	12.5	56.3	0.0	14
BooJ4.8	100.0	100.0	100.0	100.0	100.0	100.0	100.0	100.0	100.0	43.3	0.0	42.9	62.5	0.0	0.0	37.5	62.5	0.0	12
Stacking	70.0	0.0	62.5	72.7	0.0	0.0	62.5	100.0	0.0	36.7	0.0	33.3	61.5	0.0	0.0	37.5	50.0	0.0	24
CAMLET	73.3	0.0	50.0	87.5	100.0	0.0	75.0	87.5	66.7	43.3	0.0	6.7	33.3	3.3	0.0	6.7	33.3	3.3	16
GPT2																			
J4.8	90.5	66.7	85.7	100.0	0.0	100.0	100.0	91.7	0.0	76.2	0.0	66.7	90.9	0.0	0.0	100.0	83.3	0.0	5
BPNN	100.0	100.0	100.0	100.0	100.0	100.0	100.0	100.0	100.0	66.7	0.0	83.3	81.8	0.0	0.0	83.3	75.0	0.0	5
SVM	95.2	100.0	100.0	92.3	100.0	50.0	100.0	100.0	100.0	81.0	0.0	100.0	91.7	25.0	0.0	83.3	91.7	100.0	5
CLR	85.7	50.0	100.0	85.7	0.0	50.0	83.3	100.0	0.0	76.2	0.0	83.3	84.6	0.0	0.0	83.3	91.7	0.0	16
OneR	85.7	0.0	75.0	92.3	0.0	0.0	100.0	100.0	0.0	81.0	0.0	66.7	91.7	0.0	0.0	100.0	91.7	0.0	11
BagJ4.8	90.5	100.0	75.0	100.0	0.0	100.0	100.0	91.7	0.0	76.2	0.0	66.7	90.9	0.0	0.0	100.0	83.3	0.0	6
BooJ4.8	100.0	100.0	100.0	100.0	100.0	100.0	100.0	100.0	100.0	76.2	0.0	66.7	100.0	0.0	0.0	100.0	83.3	0.0	6
Stacking	61.9	66.7	0.0	100.0	0.0	100.0	0.0	91.7	0.0	71.4	0.0	83.3	76.9	0.0	0.0	83.3	83.3	0.0	11
CAMLET	81.0	0.0	75.0	84.6	0.0	0.0	100.0	91.7	0.0	76.2	0.0	28.6	47.6	0.0	0.0	28.6	47.6	0.0	8
INF1																			
J4.8	88.5	80.0	100.0	83.3	100.0	100.0	71.4	90.9	100.0	19.2	37.5	0.0	20.0	0.0	75.0	0.0	18.2	0.0	8
BPNN	100.0	100.0	100.0	100.0	100.0	100.0	100.0	100.0	100.0	26.9	40.0	22.2	25.0	25.0	50.0	28.6	18.2	25.0	6
SVM	46.2	26.7	0.0	70.0	100.0	100.0	0.0	63.6	25.0	34.6	21.4	0.0	54.5	0.0	75.0	0.0	54.5	0.0	10
CLR	53.8	100.0	0.0	47.6	66.7	50.0	0.0	90.9	50.0	19.2	33.3	0.0	28.6	0.0	25.0	0.0	36.4	0.0	16
OneR	50.0	0.0	50.0	50.0	0.0	0.0	85.7	63.6	0.0	19.2	0.0	11.1	23.5	0.0	0.0	14.3	36.4	0.0	18
BagJ4.8	96.2	80.0	100.0	100.0	100.0	100.0	100.0	90.9	100.0	26.9	33.3	37.5	22.2	0.0	50.0	42.9	18.2	0.0	10
BooJ4.8	100.0	100.0	100.0	100.0	100.0	100.0	100.0	100.0	100.0	23.1	42.9	0.0	27.3	0.0	75.0	0.0	27.3	0.0	8
Stacking	11.5	0.0	12.5	14.3	0.0	0.0	14.3	18.2	0.0	23.1	0.0	33.3	28.6	0.0	0.0	57.1	18.2	0.0	16
CAMLET	76.9	100.0	60.0	80.0	100.0	100.0	85.7	72.7	50.0	30.8	11.5	0.0	19.2	0.0	11.5	0.0	19.2	0.0	14
INF2																			
J4.8	90.6	88.2	100.0	90.9	0.0	100.0	80.0	90.9	0.0	75.0	76.5	66.7	75.0	0.0	86.7	40.0	81.8	0.0	6
BPNN	100.0	100.0	100.0	100.0	100.0	100.0	100.0	100.0	100.0	37.5	50.0	28.6	22.2	0.0	53.3	40.0	18.2	0.0	8
SVM	56.3	72.7	0.0	45.0	100.0	53.3	0.0	81.8	100.0	31.3	36.4	0.0	28.6	0.0	26.7	0.0	54.5	0.0	8
CLR	65.6	63.2	100.0	80.0	0.0	80.0	60.0	54.5	0.0	34.4	41.2	20.0	30.0	0.0	46.7	20.0	27.3	0.0	16
OneR	68.8	62.5	0.0	87.5	0.0	100.0	0.0	63.6	0.0	68.8	60.0	0.0	100.0	0.0	100.0	0.0	63.6	0.0	16
BagJ4.8	90.6	88.2	100.0	90.9	0.0	100.0	80.0	90.9	0.0	71.9	70.0	100.0	72.7	0.0	93.3	20.0	72.2	0.0	8
BooJ4.8	100.0	100.0	100.0	100.0	100.0	100.0	100.0	100.0	100.0	71.9	76.5	100.0	70.0	0.0	86.7	60.0	63.6	0.0	6
Stacking	40.6	46.2	0.0	33.3	0.0	80.0	0.0	9.1	0.0	53.1	58.8	0.0	58.3	0.0	66.7	0.0	63.6	0.0	12
CAMLET	90.6	83.3	100.0	100.0	100.0	100.0	100.0	72.7	100.0	43.8	18.8	0.0	18.8	0.0	18.8	0.0	18.8	0.0	8

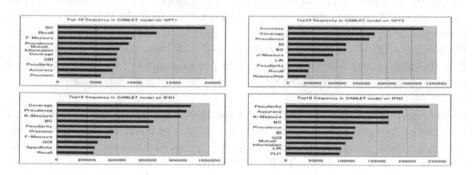

Fig. 5. Top 10 of frequencies of indices used in models of learning algorithms composed by CAMLET with 10000 bootstrap samples of the chronic hepatitis datamining results' datasets

models (Fig. 4). This shows that the medical expert evaluated these rules with both of correctness and interestingness based on his background knowledge.

On each problem, the variance of indices has been reduced in each second time datamining process. This indicates that the medical expert evaluated each second time datamining result with more certain criterion than it of each first time datamining process.

Table 6. Flow diagram to obtain datasets and the datasets of the rule sets learned from the UCI benchmark datasets

	#Mined Rules	#Class labels			%Def. class
Distribution I		L1 (0.30)	L2 (0.35)	L3 (0.35)	
anneal	95	33	39	23	41.1
audiology	149	44	58	47	38.9
autos	141	30	48	63	44.7
balance-scale	281	76	102	103	36.7
breast-cancer	122	41	34	47	38.5
breast-w	79	29	26	24	36.7
colic	61	18	19	24	39.3
credit-a	230	78	73	79	34.3
Distribution II		(0.30)	(0.65)	(0.05)	
anneal	95	26	63	6	66.3
audiology	149	49	91	9	61.1
autos	141	41	95	5	67.4
balance-scale	281	90	178	13	63.3
breast-cancer	122	42	78	2	63.9
breast-w	79	22	55	2	69.6
colic	61	22	36	3	59.0
credit-a	230	69	150	11	65.2

3.3   An Experiment on Artificial Evaluation Labels

We have also evaluated our rule evaluation model construction method with rule sets from eight datasets of UCI Machine Learning Repository [15] to investigate the performances without any human criteria.

We have taken the following eight dataset: anneal, audiology, autos, balance-scale, breast-cancer, breast-w, colic, and credit-a. With these datasets, we obtained rule sets with bagged PART, which repeatedly executes PART [8] to bootstrapped training sub-sample datasets.

To these rule sets, we calculated the 39 objective indices as attributes of each rule. As for the class of these datasets, we set up three class distributions with multinomial distribution. Table 6 shows us the process flow diagram to obtain the datasets and the description of datasets with three different class distributions. The class distribution for "Distribution I" is $P = (0.35, 0.3, 0.3)$ where p_i is the probability for class i. Thus, the number of class i in each instance D_j become $p_i D_j$. As the same way, the probability vector of "Distribution II" is $P = (0.3, 0.65, 0.05)$. We have investigated performances of learning algorithms on these balanced class distribution and unbalanced class distribution.

Constructing Proper Learning Algorithms for Rule Sets from UCI Datasets. As same as the construction of the proper learning algorithm for the meningitis data mining result, we constructed proper learning algorithms for the datasets of rule sets from the eight UCI datasets. Table7 shows an overview of constructed learning algorithms for each dataset which has two different class distributions.

Accuracy Comparison on Classification Performances. To above datasets, we have attempted the nine learning algorithms to estimate whether their classification results can go to or beyond the percentages of just predicting each default class. The left table of Table 8 shows the accuracies of the nine learning algorithms to each class distribution of the eight datasets. The learning algorithms constructed by CAMLET, boosted J4.8, bagged J4.8, J4.8 and BPNN always work better than just predicting a default class. However, their performances are suffered from probabilistic class distributions to larger datasets.

Table 7. Overview of constructed learning algorithms by CAMLET to the datasets of the rule sets learned from the UCI benchmark datasets

	Distribution I			Distribution II		
	original classifier set	overall control structure	final eval. method	original classifier set	overall control structure	final eval. method
anneal	C4.5 tree	Win+Boost+CS	Weighted Voting	C4.5 tree	Boost+CS	Weighted Voting
audiology	ID3 tree	Boost	Voting	Random Rule	Simple Iteration	Best Select.
autos	Random Rule	Win+Iteration	Weighted Voting	Random Rule	Boost	Weighted Voting
balance-scale	Random Rule	Boost	Voting	Random Rule	CS+GA	Voting
breast-cancer	Random Rule	GA+Iteration	Voting	Random Rule	Win+Iteration	Weighted Voting
breast-w	ID3 tree	Win	Weighted Voting	ID3 tree	CS+Iteration	Weighted Voting
colic	Random Rule	CS+Win	Voting	ID3 tree	Win+Iteration	Voting
credit-a	C4.5 tree	Win+Iteration	Voting	ID3 tree	CS+Boost+Iteration	Best Select.

CS means including reinfoecement of classifier set from Classifiser Systems
Boost means including methods and control structure from Boosting
Win means including methods and control structure from Window Strategy
GA means including reinforcement of classifier set with Genetic Algorithm

Table 8. Accuracies(%) on whole training datasets labeled with three different distributions(The left table). Number of minimum training sub-samples to outperform %Def. class(The right table).

Distribution I	J4.8	BPNN	SVM	CLR	OneR	Bagged J4.8	Boosted J4.8	Stacking	CAMLET
anneal	74.7	71.6	47.4	56.8	55.8	87.4	100.0	27.4	77.9
audiology	47.0	51.7	40.3	45.6	52.3	87.2	47.0	21.5	63.1
autos	66.7	63.8	46.8	46.1	56.0	89.4	66.7	29.8	53.2
balance-scale	58.0	59.4	39.5	43.4	53.0	83.3	58.0	39.5	39.5
breast-cancer	55.7	61.5	40.2	50.8	59.0	88.5	70.5	23.8	41.0
breast-w	86.1	91.1	38.0	46.8	54.4	96.2	100.0	34.2	77.2
colic	91.8	82.0	42.6	60.7	55.7	88.5	100.0	29.5	67.2
credit-a	57.4	48.7	35.7	39.1	54.8	91.3	57.4	26.5	55.7

Distribution II	J4.8	BPNN	SVM	CLR	OneR	Bagged J4.8	Boosted J4.8	Stacking	CAMLET
anneal	74.7	70.5	67.4	70.5	73.7	84.2	94.7	67.4	66.3
audiology	65.8	67.8	63.8	64.4	67.1	83.2	67.1	59.7	65.1
autos	85.1	73.8	68.1	70.2	73.8	87.9	100.0	66.7	67.4
balance-scale	70.5	69.8	64.8	65.8	69.8	80.1	85.8	62.6	63.0
breast-cancer	71.3	77.0	66.4	65.6	77.9	86.9	79.5	73.0	73.0
breast-w	74.7	86.1	73.4	68.4	74.7	87.3	100.0	63.3	70.9
colic	70.5	77.0	65.6	60.7	73.8	85.2	100.0	49.2	60.7
credit-a	70.9	70.0	65.2	65.2	71.3	85.7	87.8	61.7	65.2

Distribution I	J4.8	BPNN	SVM	CLR	OneR	Bagged J4.8	Boosted J4.8	Stacking	CAMLET
anneal	20	14	17	29	29	16	14	38	20
audiology	21	18	65	64	41	21	14	56	27
autos	38	28	76	77	70	28	28	77	31
balance-scale	12	14	15	15	32	14	9	51	128
breast-cancer	16	17	22	41	22	14	14	41	36
breast-w	7	10	10	18	14	10	6	19	11
colic	8	8	9	22	14	8	8	24	8
credit-a	9	12	16	30	28	9	8	51	19

Distribution II	J4.8	BPNN	SVM	CLR	OneR	Bagged J4.8	Boosted J4.8	Stacking	CAMLET
anneal	54	58	64	76	-	42	38	64	48
audiology	64	73	45	50	107	50	50	103	84
autos	66	102	84	121	98	45	39	78	76
balance-scale	118	103	133	162	156	86	92	132	-
breast-cancer	50	31	80	92	80	38	36	60	41
breast-w	44	36	31	48	71	34	34	52	53
colic	28	24	48	30	42	28	22	48	54
credit-a	118	159	-	-	173	76	76	120	109

Estimating Minimum Training Subset to Construct a Valid Rule Evaluation Model. As same as the case study of meningitis dataset, we have estimated the minimum training subsets for a valid model, which works better than just predicting a default class as shown in the right table in Table8. To datasets with balanced class distribution, almost of learning algorithm can construct valid models with less than 20% of given training datasets. However, to datasets with unbalanced class distribution, they need more training subsets to construct valid models, because their performances with whole training dataset fall to the percentages of default class of each dataset as shown in the left table in Table8.

4 Conclusion

In this paper, we have described the evaluation of the nine learning algorithms for a rule evaluation support method with rule evaluation models to predict evaluations for an IF-THEN rule based on objective indices, re-using evaluations of a human expert.

As the result of the performance comparison with the nine learning algorithms on the dataset of meningitis data mining result, rule evaluation models have achieved higher accuracies than just predicting each default class. To this dataset, the learning algorithm constructed by CAMLET shows higher accuracy with higher reliability than the other eight learning algorithm including three meta-learning algorithm. From the results on the datasets of hepatitis datamining results, we find out that the difference of human evaluation criteria appear as the differences of rule evaluation models on both of performances and their contents. To datasets of rule sets obtained from the eight UCI datasets, although hyper-plane type learners, such as SVM and CLR, and Stacking have failed to go to the percentage of default class of some datasets, the other learning algorithms have been able to go to or beyond each percentage of default class with smaller than 50% of each training dataset. Considering the difference between the actual evaluation labeling and the artificial evaluation labeling, it is shown that the medical expert evaluated with noticing particular relations between an antecedent and a class/another antecedent in each rule. This indicates that our approach can detect differences of human criteria as differences of performances of rule evaluation models.

As future works, we will improve the method repository of CAMLET to construct more suitable learning algorithms for rule evaluation models. We will also apply this rule evaluation support method to other datasets from various domains.

References

1. Abe, H. and Yamaguchi, T.: Constructive Meta-Learning with Machine Learning Method Repositories, in Proc. of the 17th International Conference on Industrial and Engineering Applications of Artificial Intelligence and Expert Systems IEA/AIE 2004, LNAI 3029, (2004) 502–511
2. Abe, H., Ohsaki, M., Yokoi, H., and Yamaguchi, T.: Implementing an Integrated Time-Series Data Mining Environment based on Temporal Pattern Extraction Methods – A Case Study of an Interferon Therapy Risk Mining for Chronic Hepatitis –, JSAI2005 Workshops, LNAI 4012, 425–435
3. Ali, K., Manganaris,S., Srikant, R.: Partial Classification Using Association Rules. in Proc. of Int. Conf. on Knowledge Discovery and Data Mining KDD-1997 (1997) 115–118
4. Booker, L. B., Holland, J. H., and Goldberg, D. E.: Classifier Systems and Genetic Algorithms, Artificail Inteligence, 40 (1989) 235–282
5. Breiman, L.: Bagging Predictors, Machine Learning, 24(2) (1996) 123–140
6. Brin, S., Motwani, R., Ullman, J., Tsur, S.: Dynamic itemset counting and implication rules for market basket data. Proc. of ACM SIGMOD Int. Conf. on Management of Data (1997) 255–264
7. Frank, E., Wang, Y., Inglis, S., Holmes, G., and Witten, I. H.: Using model trees for classification, Machine Learning, Vol.32, No.1 (1998) 63–76

8. Frank, E, Witten, I. H., Generating accurate rule sets without global optimization, in Proc. of the Fifteenth International Conference on Machine Learning, (1998) 144–151

9. Freund, Y., and Schapire, R. E.: Experiments with a new boosting algorithm, in Proc. of Thirteenth International Conference on Machine Learning (1996) 148–156

10. Gago, P., Bento, C.: A Metric for Selection of the Most Promising Rules. Proc. of Euro. Conf. on the Principles of Data Mining and Knowledge Discovery PKDD-1998 (1998) 19–27

11. Goodman, L. A., Kruskal, W. H.: Measures of association for cross classifications. Springer Series in Statistics, 1, Springer-Verlag (1979)

12. Gray, B., Orlowska, M. E.: CCAIIA: Clustering Categorical Attributes into Interesting Association Rules. Proc. of Pacific-Asia Conf. on Knowledge Discovery and Data Mining PAKDD-1998 (1998) 132–143

13. Hamilton, H. J., Shan, N., Ziarko, W.: Machine Learning of Credible Classifications. in Proc. of Australian Conf. on Artificial Intelligence AI-1997 (1997) 330–339

14. Hatazawa, H., Negishi, N., Suyama, A., Tsumoto, S., and Yamaguchi, T.: Knowledge Discovery Support from a Meningoencephalitis Database Using an Automatic Composition Tool for Inductive Applications, in Proc. of KDD Challenge 2000 in conjunction with PAKDD2000 (2000) 28–33

15. Hettich, S., Blake, C. L., and Merz, C. J.: UCI Repository of machine learning databases [http://www.ics.uci.edu/~mlearn/MLRepository.html], Irvine, CA: University of California, Department of Information and Computer Science, (1998).

16. Hilderman, R. J. and Hamilton, H. J.: Knowledge Discovery and Measure of Interest, Kluwe Academic Publishers (2001)

17. Hinton, G. E.: "Learning distributed representations of concepts", in Proc. of 8th Annual Conference of the Cognitive Science Society, Amherest, MA. REprinted in R.G.M.Morris (ed.) (1986)

18. Holte, R. C.: Very simple classification rules perform well on most commonly used datasets, Machine Learning, Vol. 11 (1993) 63–91

19. Klösgen, W.: Explora: A Multipattern and Multistrategy Discovery Assistant. in Fayyad, U. M., Piatetsky-Shapiro, G., Smyth, P., Uthurusamy R. (Eds.): Advances in Knowledge Discovery and Data Mining. AAAI/MIT Press, California (1996) 249–271

20. Michalski, R., Mozetic, I., Hong, J. and Lavrac, N.: The AQ15 Inductive Learning System: An Over View and Experiments, Reports of Machine Learning and Inference Laboratory, No.MLI-86-6, George Mason University (1986).

21. Mitchell, T. M.: Generalization as Search, Artificial Intelligence, 18(2) (1982) 203–226

22. Ohsaki, M., Sato, Y., Kitaguchi, S., Yokoi, H., and Yamaguchi, T.: Comparison between Objective Interestingness Measures and Real Human Interest in Medical Data Mining, in Proc. of the 17th International Conference on Industrial and Engineering Applications of Artificial Intelligence and Expert Systems IEA/AIE 2004, LNAI 3029, (2004) 1072–1081

23. Ohsaki, M., Kitaguchi, S., Kume, S., Yokoi, H., and Yamaguchi, T.: Evaluation of Rule Interestingness Measures with a Clinical Dataset on Hepatitis, in Proc. of ECML/PKDD 2004, LNAI3202 (2004) 362–373

24. Piatetsky-Shapiro, G.: Discovery, Analysis and Presentation of Strong Rules. in Piatetsky-Shapiro, G., Frawley, W. J. (eds.): Knowledge Discovery in Databases. AAAI/MIT Press (1991) 229–248

25. Platt, J.: Fast Training of Support Vector Machines using Sequential Minimal Optimization, Advances in Kernel Methods - Support Vector Learning, B. Schölkopf, C. Burges, and A. Smola, eds., MIT Press (1999) 185–208
26. Quinlan, J. R. : Induction of Decision Tree, Machine Learning, 1 (1986) 81–106
27. Quinlan, R.: C4.5: Programs for Machine Learning, Morgan Kaufmann Publishers, (1993)
28. Rijsbergen, C.: Information Retrieval, Chapter 7, Butterworths, London, (1979) http://www.dcs.gla.ac.uk/Keith/Chapter.7/Ch.7.html
29. Smyth, P., Goodman, R. M.: Rule Induction using Information Theory. in Piatetsky-Shapiro, G., Frawley, W. J. (eds.): Knowledge Discovery in Databases. AAAI/MIT Press (1991) 159–176
30. Tan, P. N., Kumar V., Srivastava, J.: Selecting the Right Interestingness Measure for Association Patterns. Proc. of Int. Conf. on Knowledge Discovery and Data Mining KDD-2002 (2002) 32–41
31. Witten, I. H and Frank, E.: DataMining: Practical Machine Learning Tools and Techniques with Java Implementations, Morgan Kaufmann, (2000)
32. Wolpert, D. : Stacked Generalization, Neural Network 5(2) (1992) 241–260
33. Yao, Y. Y. Zhong, N.: An Analysis of Quantitative Measures Associated with Rules. Proc. of Pacific-Asia Conf. on Knowledge Discovery and Data Mining PAKDD-1999 (1999) 479–488
34. Zhong, N., Yao, Y. Y., Ohshima, M.: Peculiarity Oriented Multi-Database Mining. IEEE Trans. on Knowledge and Data Engineering, 15, 4, (2003) 952–960

Training Classifiers for Unbalanced Distribution and Cost-Sensitive Domains with ROC Analysis

Xiaolong Zhang, Chuan Jiang, and Ming-jian Luo

School of Computer Science and Technology,
Wuhan University of Science and Technology, Wuhan 430081 China
{xiaolong.zhang, chuan.jiang, mingjian.luo}@mail.wust.edu.cn

Abstract. ROC (Receiver Operating Characteristic) has been used as a tool for the analysis and evaluation of two-class classifiers, even the training data embraces unbalanced class distribution and cost-sensitiveness. However, ROC has not been effectively extended to evaluate multi-class classifiers. In this paper, we proposed an effective way to deal with multi-class learning with ROC analysis. An EMAUC algorithm is implemented to transform a multi-class training set into several two-class training sets. Classification is carried out with these two-class training sets. Empirical results demonstrate that the classifiers trained with the proposed algorithm have competitive performance for unbalanced distribution and cost-sensitive domains.

Keywords: Classification, ROC, Cost-Sensitive Learning, Error Correcting Output Coding.

1 Introduction

Classification learning is usually measured by accuracy. However, when skewed class distribution or unequal classification error cost happens, accuracy could not ensure the total error cost of the classification algorithm to be minimum. For example, in the industrial risk management like the medical decision, among the healthcare data process, the "healthy" cases are far more than "patient" ones; when checking the abuse of credit cards, the "normal use" cases are far more than the "abuse use" ones. Moreover, in some domains the data class's distribution can vary remarkably. For example, the proportion of finance fraud varied significantly from month to month and place to place [1]. In these applications, the accuracy to predict all the samples to be "healthy" or "normal" can be more than 0.99, since more than 99% data belong to "healthy" or "normal". Therefore, in classification learning of the unbalanced distribution data classes, accurate rate 0.99 cannot indicate a classifier is of a good classification ability. In such a domain, accuracy does not have the ability to evaluate the learning results or learning performance. Traditional machine learning techniques basically consider the case that the algorithms learn with the balanced data, where it is supposed the training samples distribution is uniform and classification error cost (i.e., misclassification of a positive example is equal to misclassification

A. Hoffmann et al. (Eds.): PKAW 2006, LNAI 4303, pp. 89–98, 2006.

of negative example) is equal both in the predicting and training data. In fact, the practical application data is always unbalanced, especially in medical diagnoses, pattern recognition, decision-making theory, as well as most kind of risk management domains.

ROC (Receiver Operating Characteristic) analysis [2] is an evaluation tool. By using the graphical mode, ROC can be used to measure the classification ability of a classifier in the conditions of any data distribution and error cost. ROC curve has less sensitivity of the class distribution and error cost, which makes ROC be a useful evaluation criteria for learning cost-sensitive or/and unbalanced class data. ROC analysis is ready to use for two-class classification domains. However, the methods to classify the multi-class problems need further studying.

This paper introduces an algorithm EMAUC which is based on ROC analysis, to do classification of multi-class domains. This algorithm is by means of Error Correcting Output Codes [3] to transform a multi-class dataset to several two-class datasets, generate ROC curves from these two-class datasets, and finally synthesize a multi-class classifier. Compared with other multi-class classifiers, EMAUC has the advantage of competitive performance, better comprehension and no sensitivity of skew datasets. EMAUC algorithm employs two-class classifiers to finish two-class classification. It is based on a machine learning tool (WEKA [4]) and ROC analysis graph tool (ROCon [5]). Experimental results with UCI [6] datasets show that EMAUC has a good learning performance in dealing with multi-class classification domains (including unbalanced distribution and cost-sensitive domains).

The rest of this paper is arranged as following. Section 2 describes ROC analysis. Section 3 presents EUMAC algorithm. Section 4 demonstrates the experimental results. Section 5 is the related work in this area. The final section is the conclusion.

2   ROC Analysis

ROC analysis origins from the statistical decision-making theory in the 1950s and has been used to introduce the connection with hit rates and false alarm rates of classifiers. ROC has been used in the domains such as transistor, psychology, medical decision-making and so on. Swets [7] extends the ROC analysis to wider application domains. ROC analysis has the ability to evaluate the classifiers for learning with unbalanced class distribution and unequal classification error cost. As mentioned above, accuracy is difficult to evaluate the learning results with unbalanced distribution and unequal classification error cost, ROC may replace the accuracy to be a better evaluation criterion.

As we know that the accuracy of a two-class classifier can be described with a 2×2 confusion matrix. Assume that the ratio of negatives incorrectly classified to total negatives is FP rate, and the ratio of positives correctly classified to total positives is TP rate, ROC can be described by a two-dimension graph [8] in which TP rate is plotted on the Y axis and FP rate is plotted on the X axis.

On one hand, the classifiers such as C4.5 and SVM whose output results are some classes can produce a pair of TP and FP in the dataset; On the other hand, the classifiers such as Native Bayes and Neural Network whose output results are some numerical values which indicate the possibility for a training example to belong to a class. By setting a threshold, if this possibility is higher than the threshold, the class of this example can be transformed to "Yes" class, otherwise "No".

There are wide applications of ROC with two-class classifiers. The evaluation criterion has been developed into Cost Curve [9], AUC(Area Under the ROC) [20] and so on. AUC has been used as a criterion to evaluate the learning performance, since it simply describes the integrative capability by calculating the area under the ROC convex hull. It is difficult to describe the learning performance of multi-class classifiers. According to the two-class confusion, there is a $N \times N$ confusion matrix including N correct classification (elements in the positive diagonal) and $N^2 - N$ error classification (elements beyond the positive diagonal). Within the multi-class ROC analysis, the relationship of TP and FP is of $N^2 - N$ independent variables, where there exists a $N^2 - N$ dimensional space. For example, given a three-class learning domain, the points in the ROC space become a $(3^2 - 3) = 6$ dimensional polytope. In fact, how to calculate the convex hull of super-geometrical object is a NP difficult problem [8]. Therefore, the solution for multi-class classifier based on ROC analysis cannot be found by directly extending the technique of two-class ROC analysis.

There is some work to extend two-class classification to multi-class classification in the ROC analysis. The classifiers based on three-class ROC analysis [8,11,14] are difficult to extend to more than five classes because of the computational feasibility. Therefore, researchers find other methods like OVA (One-Vs-All), Pairwise to avoid the computational complexity. But they cannot directly select the best algorithm used in the learning like the two-class ROC analysis. The work of HTM method [15] aims to solve the multi-class problems, HTM uses a function to compute the average pairwise AUCs in multi-class classifications which is nearly like the pairwise method by extending AUC. It acquires the best multi-class classification measuring performance without considering any misclassification cost. Therefore there are still some issues in HTM method.

3   EMAUC Algorithm

This section describes a method to solve multi-class ROC problem with an algorithm EMAUC (Multi-AUC with ECOC). By the ECOC (Error Correcting Output Codes) [3], EMAUC transforms a multi-class dataset to several two-class datasets. By coding the target classes, ECOC transforms the multi-class training set to several independent two-class training sets. The transformation can be explained with the following example.

Table 1 shows a data set with three target class values "S","V","I". Table 2 is a kind of ECOC form. Each column will be used to generate a two-class training data. For instance, the first column binary code (1 0 0) is employed to replace

Table 1. A Multi-class Data Set

Sepal Length	Sepal Width	Petal Length	Petal Width	Class
4.6	3.2	1.4	0.2	S
5.3	3.7	1.5	0.2	S
7.0	3.2	4.7	1.4	V
6.4	3.2	4.5	1.5	V
6.3	3.3	6.0	2.5	I
5.8	2.7	5.1	1.9	I

Table 2. The ECOC for Multi-class Data Transformation

Class	Code Cluster
S	1 1 1
V	0 0 1
I	0 1 0

Table 3. A Transformed Data Set with ECOC (1)

Sepal Length	Sepal Width	Petal Length	Petal Width	Class
4.6	3.2	1.4	0.2	1
5.3	3.7	1.5	0.2	1
7.0	3.2	4.7	1.4	0
6.4	3.2	4.5	1.5	0
6.3	3.3	6.0	2.5	0
5.8	2.7	5.1	1.9	0

Table 4. A Transformed Data Set with ECOC (2)

Sepal Length	Sepal Width	Petal Length	Petal Width	Class
4.6	3.2	1.4	0.2	1
5.3	3.7	1.5	0.2	1
7.0	3.2	4.7	1.4	0
6.4	3.2	4.5	1.5	0
6.3	3.3	6.0	2.5	1
5.8	2.7	5.1	1.9	1

Table 5. A Transformed Data Set with ECOC (3)

Sepal Length	Sepal Width	Petal Length	Petal Width	Class
4.6	3.2	1.4	0.2	1
5.3	3.7	1.5	0.2	1
7.0	3.2	4.7	1.4	1
6.4	3.2	4.5	1.5	1
6.3	3.3	6.0	2.5	0
5.8	2.7	5.1	1.9	0

Input: Training Set S, Two-classifier C, Code Type T
Output: The value of EMAUC
    Begin
        Initialize(S);
        Code = GenerateCorrectCode(S,T);
        For each column of Code (i)
            SubTrainingSet(i) = FilterTrainingSet(S, Code(i));
            Classifier(i) = Construct Classifier(SubTrainingSet(i),C);
        EndFor;
        EMAUC = Avg(AUC(i));
    End.

Fig. 1. EMAUC Algorithm

the corresponding target class (S V I) of Table 1, respectively. Table 3 is a two-class training set that is the result of the replacement of the target class with the code listed in the first column of Table 2. For the other two columns in Table 2, EMAUC generates two other two-class training sets (see Tables 4, and 5).

In fact, there are several coding methods, i.e., exhausting code, nature code, random code [3]. Different coding method generates different code length and different separating ability. According to the result of [3], ECOC is selected as the coding method in our algorithm.

Compared with the OVA and Pairwise, the two-class datasets created by ECOC can be independent each other. It makes EMAUC possible to deal with multi-class problems. EMAUC learns every two-class dataset using a selected two-class classifier, and acquires related two-class confusion matrices. After the AUC of each two-class classification is computed, a one-dimensional vector can be obtained. This vector will be used to evaluate the multi-class classification algorithm. In this paper, the value of EMAUC is the arithmetic mean of the elements in the vector.

Fig. 1 describes the outline of EMAUC algorithm. The input of the algorithm includes a multi-class training set S, a two-class classifier (such as Native Bayes, Logistic) and a code type T (as mentioned above, ECOC code is selected in this algorithm). The output of the algorithm is the value of EMAUC.

$Initialize(S)$ does initial work such as reading data, analyzing attribute and class number, collecting dataset characteristics. Given a code type T (ECOC code type) and a multi-class training set S, the algorithm generates code words. According to the code words of code table to replace the class in the multi-class training set S, it produces several two-class training sets. Based on these two-class training sets, two-class classifier can be built, where Native Bayes, Logistic can be used as such a classifier. After a two-class classifier is trained, the algorithm calculates the AUC of each two-class classifier. Finally, the average of these AUCs is the output of the algorithm EMAUC.

EMAUC is implemented based on WEKA [4] platform. The AUC is computed by means of ROCon graph tool [5]. WEKA platform is developed by Waikato University in New Zealand. It includes variety of algorithms, like classification, cluster, associate rule, regression, and visualization. The source code (Java program) of WEKA platform is open. Therefore one can rewrite the source code and improve the capability of some algorithms. EMAUC trains two-class training sets with C4.5 [16], Logistic, Native Bayes, NBtree and NN (Neural Network). Finally, it calculates the AUC with a reconstructive ROCon tool.

Table 6. Datasets Used in EMAUC

UCI Dataset	Dataset Name	Sample Number	ATTRIBUTE NUMBER			Class Number
			Total	Discrete	Continuity	
anneal	ANN	898	38	32	6	6
artifical	ART	5109	7	0	7	10
audiology	AUD	226	69	69	0	24
auto-mpg	AUT	399	7	2	5	4
autos	AUS	205	25	10	15	7
balance-scale	BAL	626	4	0	4	3
bridges2	BRI	108	11	10	1	6
flag	FLA	194	27	17	10	6
glass	GLA	214	9	0	9	7
hayes-roth	HEY	132	4	4	0	3
heart-c	HEA	303	13	7	6	5
heart-h	HEH	294	13	7	6	5
hypothyroid	HYP	3772	29	22	7	4
lymph	LYM	148	18	15	3	4
machine	MAC	209	7	0	7	8
page-blocks	PAG	5473	10	0	10	5
primary-tumor	PRI	339	17	17	0	22
segment	SEG	2310	19	0	19	7
solar-flare	SOL	333	10	10	0	8
soybean	SOY	683	35	35	0	19
vehicle	VEH	848	18	0	18	4
waveform-5000	WF5	5004	40	0	40	3
vowel	VOW	990	13	3	10	11
wine	WIN	178	13	0	13	3
zoo	ZOO	104	17	16	1	7

4   Experimental Results

The experiments have been carried out within 25 UCI [6] datasets. Table 6 describes these datasets, whose number of target number is more than 3. Our system transforms a multi-class training set into several two-class training sets. For each two-class training set, EMAUC invokes C4.5, Logistic, Native Bayes, NBtree and NN to train two-class classifiers respectively. Then the system computes the AUC value for each two-class training set. Therefore, the EMAUC of each multi-class training set can be acquired with these AUCs.

Table 7 demonstrates the learning performance evaluated by EMAUC with the 25 datasets, where the two-class classifiers are ZeroR, C4.5, Logistics, Native

Table 7. EMAUC Derived from Different Two-class Classifiers

Dataset	ZeroR	C4.5	Logistic	Native Bayes	NBtree	NN
ANN	0.500	0.916	0.915	0.891	0.914	0.998
ART	0.500	0.688	0.915	0.676	0.862	0.820
AUD	0.500	0.998	0.926	0.941	0.981	0.989
AUT	0.500	0.882	0.985	0.826	0.969	0.979
AUS	0.500	0.995	0.918	0.882	0.995	0.994
BAL	0.500	0.722	0.995	0.741	0.650	0.809
BRI	0.500	0.953	0.660	0.923	0.965	0.986
FLA	0.500	0.989	0.959	0.912	0.984	0.989
GLA	0.500	0.839	0.973	0.807	0.900	0.912
HAY	0.500	0.986	0.928	0.986	0.986	0.990
HEA	0.500	0.811	0.969	0.800	0.877	0.889
HEH	0.500	0.770	0.840	0.757	0.827	0.840
HYP	0.500	0.972	0.765	0.958	0.972	0.892
LYM	0.500	1.000	0.811	0.981	0.998	0.949
MAC	0.500	0.934	0.995	0.934	0.949	0.991
PAG	0.500	0.973	0.957	0.954	0.991	0.976
PRI	0.500	0.831	0.990	0.814	0.877	0.976
SEG	0.500	0.966	0.906	0.889	0.999	0.998
SOL	0.500	0.883	0.999	0.823	0.871	0.886
SOY	0.500	0.999	0.605	0.954	0.996	0.999
VEH	0.500	0.903	0.998	0.749	0.907	0.974
WF5	0.500	0.813	0.981	0.825	0.998	0.998
VOW	0.500	0.953	0.995	0.926	0.974	0.993
WIN	0.500	1.000	0.996	0.999	0.995	1.000
ZOO	0.500	1.000	0.994	0.986	0.999	1.000
Average	0.500	0.911	0.919	0.877	0.937	0.953

Bayes, NBtree, NN (Neural Network). ZeroR is used as dumb classifier, whose AUC value is 0.500 for any training. The average accuracy of these two-class classifiers are 0.500, 0.911, 0.919, 0.877, 0.937, 0.953, respectively. Among these 5 two-class classifiers, Neural Network has the best learning performance. Logistics and C4.5 have almost the same performance. Both of Logistics and C4.5 are better than Native Bayes. C4.5 algorithm is better than Logistic in 12 datasets but worse than Logistics in other 13 datasets. The average value of NBtree algorithm is 0.937 which is better than Native Bayes and C4.5.

As Dietterich [3] points out that ECOC improves the probability estimate, this basically accords with our experimentation. From Table 7 we also know NN has the best performance among these 5 classification algorithms, due to its average value is 0.953 over the 25 datasets. Although the average capability of NN is the best among these 5 classification algorithms, in some domains, NN may not have the best performance. As the results given by Table 7, C4.5 is the best in 8 datasets, Logistic is the best in 11 datasets, NBtree is the best in 5 dataset, and Native Bayes has none. It seems that the best application range of each classification algorithm is nearly correlative with the data characteristic of a given training dataset.

5   Related Work

ROC analysis has been used as a useful tool for machine learning because it allows to assess the decision functions which cannot be calibrated, even when the prior distribution of the classes is unknown [17]. Provost and Fawcett [19] discuss the evaluation of learning algorithms by use of ROC other than accuracy. Ling et al. [20] have proved AUC is a statistically consistent and more discriminating measurement than accuracy. This research group also has experimentally compared the performance of the NB, C4.5 and SVM with AUC in [21], but only uses some two-class training sets for evaluation. These works are not considered about how to divided a multi-class training set to two-class training sets, which is different from our work. With ECOC, we separate a multi-class data to two-class data sets.

Flach and Lachiche [22] use ROC curves to improve accuracy and cost of multi-class probabilistic classifiers. They use a hill-climbing approach to adjust the weights of each class. A multi-class probabilistic classifier is turned into a categorical classifiers by setting weights on the class scores for all classes and assigning the class which maximizes the weighted score. Then they find the best weight by heightening score to optimal cost or accuracy. This method is also different from ours in training style. Our algorithm first divides a given multi-class data set to several two-class training sets with ECOC, then train each two-class training set, and finally computes the EMAUC from these training results.

By now, AUC is not only used to evaluate the classification algorithms, also used to maximize variants of learning methods including SVM, boosting, regression and so on.

6   Conclusion

This paper introduces a multi-class ROC analysis technique based on ECOC. This method has been implemented and experimentally evaluated with a set of multi-class training sets. The results demonstrate that this technique has competitive performance, better comprehension and less sensitivity with skew datasets. The proposed algorithm takes the average AUC value of each two-class classifier to calculate EMAUC. One of the future works is to find some applications of the proposed algorithm EMAUC in the real world domains.

Acknowledgements

We thank the members of Machine Learning and Artificial Intelligence Laboratory, School of Computer Science and Technology, Wuhan University of Science and Technology, for their helpful discussion within seminars. This work was supported in part by the Scientific Research Foundation for the Returned Overseas Chinese Scholars, State Education Ministry, and the Project (No.2004D006) from Hubei Provincial Department of Education, P. R. China.

References

1. T. Fawcett and F. Provost. Adaptive Fraud Detection. *Data Mining and Knowledge Discovery*, 291-316, 1997.
2. L. B. Lusted. Logical Analysis in Roentgen Diagnosis. *Radiology*. 74:178-193. 1960.
3. T.G.Dietterich, G.Bakiri. Solving Multiclass Learning Problems Via Error Correcting Output Codes. *Journal of Artificial Intelligence Research*, 2:263-286, 1995.
4. WEKA. www.cs.waikato.ac.nz/ml/weka.
5. ROCon. www.cs.bris.ac.uk/Research/MachineLearning/rocon.
6. C.J. Merz, P.M. Murphy, and D.W. Aha. UCI repository of machine learning databases, University of California, Irvine. Available: http://www.ics.uci.edu/ mlearn/MLRepository.html, 1998.
7. J. A. Swets, R. M. Dawes, and J. Monahan. Better Decisions through Science. *Scientific American*, 2000.
8. T. Fawcett. ROC Graphs: Notes and Practical Considerations for Researchers. *Machine Learning*, 2004.
9. C. Drummond and R. C. Holte. What ROC Curves Can't Do (and Cost Curves Can). *Proceedings of the ROC Analysis in Artificial Intelligence, 1st International Workshop*, 19-26, 2004.
10. C. X. Ling, J. Huang, and H. Zhang. AUC: a Better Measure than Accuracy in Comparing Learning Algorithms. *Canadian Conference on AI*, 2003.
11. D.Mossman. Three-way ROCs. *Medical Decision Making*, 19(1):78-89. Srinivasan, 1999.
12. C. Ferri, P. A. Flach, and J. Hernandez-Orallo. Learning Decision Trees Using the Area Under the ROC Curve. *In Proceedings of the Nineteenth International Conference on Machine Learning ICML*, 139-146, 2002.
13. C. Ferri, J. Hernndez-Orallo,and M.A. Salido. Volume Under the ROC Surface for Multi-class Problems. *Proceedings of 14th European Conference on Machine Learning, ECML*, 2003.
14. C. Ferri, J. Hernndez-Orallo, and M.A. Salido. Volume Under the ROC Surface for Multi-class Problems. Exact Computation and Evaluation of Approximations. 2003, Univ. Politecnica de Valencia: Valencia. 1-40. DSIC. Univ. Politc. Valncia. 2003.
15. D.J. Hand and R.J. Till. A Simple Generalization of the Area Under the ROC Curve for Multiple Class Classification Problems. *Machine Learning*, 45(2): 171-186, 2001.
16. J.R. Quinlan. C4.5: Programs for Machine Learning. *San Mateo, California: Morgan Kaufmann*, 1993.
17. A.P. Bradley. The Use of the Area under the ROC Curve in the Evaluation of Machine Learning Algorithms. *Pattern Recognition*, 30:1145-1159.1997.
18. P.A. Flach. The Geometry of ROC Space: Using ROC Isometrics to Understand Machine Learning Metrics. *Proceedings of the International Conference on Machine Learning*, 2003.

19. F. J. Provost and T. Fawcett. Analysis and Visualization of Classifier Performance: Comparison under Imprecise Class and Cost Distributions. *Knowledge Discovery and Data Mining*, 43-48, 1997.
20. C.X. Ling, J. Huang, and H. Zhang. AUC: a Statistically Consistent and More Discriminating Measure Than Accuracy. *Proceedings of 18th International Conference on Artificial Intelligence (IJCAI-2003)*, 329-341, 2003.
21. J.Huang, J.Lu, and C.X. Ling. Comparing Natives Bayes, Decision Trees, and SVM using Accuracy and AUC. *Proceedings of European Conference on Data Mining (ICDML-2003)*, 2003.
22. N. Lachicle and P. Flach. Improving Accuracy and Cost of Two-Class and Multi-Class Probabilistic Classifiers Using ROC Curves. *Proceedings of the Twentieth International Conference on Machine Learning (ICML-2003)*, 2003.

Revealing Themes and Trends in the Knowledge Domain's Intellectual Structure

Tsung Teng Chen[1] and Maria R. Lee[2]

[1] National Taipei University, Graduate School of Information Management
104 Taipei, Taiwan
misttc@mail.ntpu.edu.tw
[2] Shih Chien University, Department of Information Management
104 Taipei, Taiwan
maria.lee@mail.usc.edu.tw

Abstract. Thousands of academic papers related to scientific research appear every year, and the accumulated literatures over the years are voluminous. It may be very helpful to be able to visualize the entire body of scientific knowledge, and to be able to track the latest developments in the science and technology fields. However, to make knowledge visualizations clear and easy to interpret are very challenging tasks. This paper draws on a vast amount of knowledge management related citations to reveal the main themes and trends of this particular research field. It aims to help researchers to grasp the research focus and trends in knowledge management, which are hard to discern due to the sheer amount of relevant literatures. It also helps novice researchers gain useful and instant insights into the most important literatures in their chosen field of research.

1 Introduction

Thousands of academic papers related to scientific research appear every year, and the accumulated literatures over the years are voluminous. It is may be very useful be able to visualize the entire body of scientific knowledge and track the latest developments in particular science and technology fields. However, to make knowledge visualizations clear and easy to interpret are challenging tasks. Many studies have drawn their citation data by using a key phrase to query citation indexes. Retrieving citation data by a simple query of citation indexes is a rather crude and limiting technique.

A knowledge domain is represented collectively by research papers and their inter-relationships in this research area. A knowledge domain's intellectual structure can be discerned by studying the citation relationships and analyzing seminal literatures of that knowledge domain. Knowledge Management (KM) is a fast growing field with great potential. However, researchers have disagreeing opinions about what constitutes the content and context of the KM research area [1]. Our Intellectual Structure of the KM domain has been constructed with predominantly information systems and management oriented factors [2, 3]. Our study drew primarily on voluminous science and engineering literature that has given us some interesting

A. Hoffmann et al. (Eds.): PKAW 2006, LNAI 4303, pp. 99–107, 2006.

results. The methodology we used to build the intellectual structure of the KM domain is reviewed in section 2 of this paper. Sections 3 and 4 present the KM structure and its research trends respectively.

2 Background Analysis in Knowledge Domain Intellectual Structure

The study of the intellectual structure of a discipline was pioneered by researchers in information science in the early eighties [4]. The Author Co-citation Analysis (ACA), a bibliometric technique, is used in presenting the intellectual structures of a knowledge domain. Recent studies in knowledge visualization adapt this ACA approach as its underlying methodology and outfit the intellectual structure with visual cues and effects [5-8]. The ACA intellectual structure method has been applied in much research [3, 9-11]. Our proposed approach is similar to ACA analysis to derive the intellectual structure of the domain of KM.

2.1 The ACA Knowledge Intellectual Structure Methodology

Authors who have made influential contributions to a given discipline constitute the unit of analysis in ACA. An author represents a reference to the concept for which the author is recognized by their works and citations received by their works. Co-citation analysis thus infers the relationship between key concepts, which is represented by the extent of joint citations of authors making seminal contributions in any given field. The approach relies on the intuitionally appealing thought that authors contributing to concepts viewed as being overlapping or closely related are more likely to be cited together by other researchers than by authors contributing to concepts viewed as distinct or different [3]. The method starts with identifying groups of authors who are frequency co-cited; they are grouped together based on their co-citations. The groups of authors are usually compiled from a list of the most cited authors using a representative key phrase, such as Knowledge Management, querying papers published between a designated period included in the Social Science Citation Index, Science Citation Index, or some other citation index. The co-citation relationships between authors are usually represented by a co-citation matrix, which serves as the input of factor analysis. The intellectual structure or subfields represented by factors derives from the factor analysis.

As an example of how ACA works, Chen [8] studied author co-citation patterns found in IEEE Computer Graphics and Applications (CG&A) magazine for a period of 18 years. The resulting CG&A citation data included 10,292 unique articles written by 5,312 authors. Those authors who had received less than five citations in CG&A were filtered out, and only 353 most cited authors were left for further analysis. Sixty factors were established from the factor analysis applied on the author co-citation matrix. Five most significant factors that explained 39% of the variance were discussed in detail by listing the contributing authors of each factor.

2.2 The Intellectual Structure Layout Methods

The co-citation matrix computes a correlation matrix of Pearson correlation coefficients. Researchers [8, 12] have applied Pathfinder network (PFNET) scaling to prune the network that the correlation matrix defines. PFNET scaling is used to extract the most important relationships from the correlation matrix [13]. The topology of a PFNET is determined by two parameters q and r and the corresponding network is denoted as PFNET(r, q). The parameter q constrains the scope of minimum cost paths to be calculated. The parameter r defines the Minkowski metric used for computing the distance of a path. The result of PFNET scaling is shown spatially as a sparsely connected graph whereas authors are represented by nodes in the graph. Other layouts, such as the multidimensional scaling map (MDS), which uses the correlation matrix to provide a spatial representation of authors have also been utilized [3].

2.3 Related Works Applying the Methodology in Knowledge Management

Ponzi [2] studied the intellectual structure and interdisciplinary breadth of KM. Intellectual structure is established by a principal component analysis applied to an author co-citation frequency matrix. The author co-citation frequencies used were derived from the 1994-1998 academic literature and captured by the single search phrase of "Knowledge Management." The study found four factors, which were labeled Knowledge Management, Organizational Learning, Knowledge-based Theories, and The Role of Tacit Knowledge in Organizations. The interdisciplinary breadth surrounding Knowledge Management was discovered mainly in the discipline of management. The study validated the hypothesis with empirical evidence that the discipline of Computer Science is not a key contributor in KM.

Subramani et al. [3] examined KM research from 1990-2002, and this examination highlighted the intellectual structure of management related researches in the field. The results revealed the existence of eight subfields of research on the topic, which include Knowledge as Firm Capability, Organizational Information Processing and IT Support for KM, Knowledge Communication, Transfer and Replication, Situated Learning and Communities of Practice, Practice of Knowledge Management, Innovation and Change, Philosophy of Knowledge, and Organizational Learning and Learning Organizations. These sub-fields reflect the influence of a wide array of foundational disciplines such as management, philosophy, and economics.

3   The Intellectual Structure of Knowledge Management

The studies reviewed above drew their citation data by using a key phrase querying citation indexes. The citation data retrieved by a simple query of citation indexes were rather limited. Our proposed approach is based on a scheme, which constructs a full citation graph from the data drawn from the online citation database CiteSeer [14]. The proposed procedure leverages the CiteSeer citation index by using key phrases to query the index and retrieve all matching documents from it. The documents retrieved by the query are then used as the initial seed set to retrieve papers that are citing or cited by literatures in the initial seed set [15]. The full citation graph is built by linking all articles retrieved, which includes more documents than the other schemes reviewed earlier. The resulted citation graph was built from the literatures and citation information retrieved by querying the term "Knowledge Management" from CiteSeer on March, 2006. The complete citation graph contains 599,692 document nodes and 1,701,081 citation arcs. In order to keep the highly cited papers and keep the literature to a manageable size, we pruned out papers were cited less than 150 times. The resultant citation graph contains 255 papers and 776 citation arcs.

3.1 Factor Analysis

The co-citation matrix is derived from the citation graph and fed to factor analysis. Seventeen factors were determined which explained 45.5% total variances. The unit of analysis used here is based on documents instead of authors. An author is considered as the proxy of the specialty S/He represents. However, a researcher's specialty may

Table 1. Factors and Loading

Factor	Descriptive Name	Variance Explained	Cumulative Variance
1	Query on Semi-structured Data	5.664	5.664
2	Inductive Learning and Inductive Logic programming	3.984	9.647
3	Efficient Search of Multi-dimensional objects	3.775	13.422
4	Logic Programming and Deductive Databases	3.586	17.008
5	Machine Learning and Classifiers	3.286	20.294
6	Distributed Problem Solvers and Rational Agents	3.018	23.313
7	Knowledge Interchange Format and Knowledge Sharing	2.672	25.985
8	Data Mining – Efficient Mining of Association Rules	2.423	28.407
9	Constraint Query Languages and Logic Programming	2.315	30.722
10	Unified Views of Information Sources	2.147	32.869
11	Modal and Temporal Logic	2.087	34.956
12	Parallel Distributed Model and Language	2.015	36.972
13	Functional Languages and Development Environment	1.899	38.870
14	Agent Development and Communication Languages	1.767	40.637
15	STRIPS Planning	1.685	42.322
16	World Wide Webs and Search Engines	1.648	43.971
17	Bayesian Networks Learning	1.549	45.520

change or evolve over time. We therefore took the document as the analysis unit in our study. The factors and their loading are listed in Table 1.

Factor one represents research on query of semi-structured data or heterogeneous Information Sources. In contrast with traditional relational and object-oriented database systems that force all data to adhere to an explicitly specified schema, much of the information available on-line, such as a WWW site, is semi-structured. Semi-structured data is relatively varied, irregular, or mutable to easily map to a fixed schema. Knowledge, in contrast with data and information, is inherently unstructured. The study of querying semi-structured data could be considered as the harbinger of the query of the semi-structured or unstructured knowledge base.

Factor two represents research on inductive learning and inductive logic programming. Inductive learning includes works focused on learning concepts from examples and learning algorithms. Inductive Logic Programming (ILP) is a machine learning approach that uses techniques of logic programming. ILP systems develop predicate descriptions from examples and background knowledge. The derived predicate descriptions, examples, and background knowledge are all described as logic programs. These areas of studies try to derive new knowledge from existing facts and background knowledge.

Factor three is characterized by researches on efficient search and data structure of multi-dimensional objects. This area of study includes the research of querying image content by specifying color, texture, and shape of image object, and the application of R+-Tree for dynamic indexing of multi-dimensional objects. Novel applications of this line of research in similarity search in massive DNA sequence databases are a new trend.

Factor four includes earlier knowledge-based related research such as that of deductive databases and logic programming. Non-monotonic reasoning and logic as well as semantical issues on non-monotonic logic and predicate logic were also covered.

Factor five represents machine learning related research, which encompasses inductive learning, statistical learning, classifiers, and learning algorithms and programs. The machine learning programs referred to here are based on a classification model that discovers and analyzes patterns found in records. C4.5 is an example of an inductive method that finds generalized rules by identifying patterns in data. This factor appears to be the transitional works that bridge or transcend from the area of machine learning to data mining.

Factor six represents distributed problem solvers and rational agents, where an agent is essentially a delegate who solves problems with human like intelligence. The study of the architecture of a resource-bounded rational agent and the formulation of the agent's intention seems to be the precursor of research in the area of intelligent agent. An intelligent agent is described as a self-contained, autonomous software module that could perform certain tasks on behalf of its users. It could also interact with other intelligent agents and/or humans in performing its task(s).

Factor seven represents Knowledge Interchange Format (KIF) and knowledge sharing. KIF is essentially a language designed for use in the interchange of knowledge among disparate computer systems. When a computer system needs to communicate with another computer system, it maps its internal data structures into KIF. Alternatively, when a computer system reads a knowledge base coded in KIF, it converts the data into its own internal form. The research of knowledge sharing tried to find ways of preserving existing knowledge bases and of sharing, reusing, and building on them. This line of research tries to develop the enabling technologies to facilitate reusing knowledge bases that have been built and used by AI systems.

Factor eight represents data mining research that discusses the effective and efficient algorithms of mining association rules from large databases. Data mining, in essence, is a knowledge discovering technique that tries to learn rules and patterns from a large amount of data. The roots of data mining can be traced back to machine learning as already mentioned.

Factor nine represents constraint query languages and constraint logic programming, which belong to the subfields of constraint programming that describe computations by specifying how these computations are constrained. Therefore, the constraint programming paradigms are inherently declarative, such as Prolog clauses.

Factor ten represents the study of providing unified views of information from diverse sources or data located within distributed and heterogeneous databases. Papers under this factor try to utilize the semantic model of a problem domain that could provide a unified view to integrate information from disparate information sources.

Factors eleven through seventeen explain less than eleven percent of the total variance. We therefore briefly review these factors altogether. Factor eleven is characterized by modal and temporal logic. Factor twelve represents a language independent model for parallel and distributed programming, such as the Linda system [16]. Factor thirteen represents functional languages and their development environments. Factor fourteen represents languages for the development and communication of information agents. Factor fifteen represents STRIPS planning, which is a simple and compact method of expressing planning problems. Instead of having complex logic statements in the knowledge base, STRIPS allows only simple positive facts and everything not explicitly listed as true in the knowledge base is considered false. Factor sixteen is characterized by World Wide Webs and search engines. Factor seventeen represents Bayesian networks, which combines knowledge with statistical data for learning.

3.2 Pathfinder Network

The Pearson correlation coefficients between items (papers) were calculated when factors analysis was applied. The correlation coefficients are used as the basis for PFNET scaling [17]. The value of Pearson correlation coefficient falls between the range -1 and 1. The coefficient approaches to one when two items correlate completely. Items that closely relate, i.e., are highly correlated, should be placed closely together spatially. The distance between nodes is normalized by taking $d = 1/(1 + r)$, whereas r is the correlation coefficient. The distance between items is inversely proportional to the correlationcoefficient, which maps less correlated items apart and highly correlated items spatially adjacent.

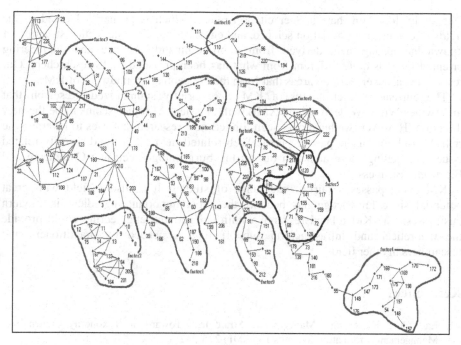

Fig. 1. PFNET Scaling with Papers under Same Factors Close to Each Other

4   Research Trends in KM

Similar to the reviewed methodologies, we sought to identify research trends in KM. Instead of searching all papers in the citation index database, we limited our search to literature published during the last four years. We do not repeat the description of the analytical procedure because it is the same as above in section 2.1. Ten factors were identified by the factor analysis procedure.

Semantic Web, Ontology, and Web Ontology related researches are recent popular research trends in the KM domain area. Distributed knowledge representation, reason systems, and description logic are also research interests due to World Wide Web proliferation. In addition, classifiers and patterns learning, especially in the area of Webs and hidden databases with Web front end, are active research areas too. Generally speaking, Extensible Markup Language (XML) and related topics are increasing. The issue of trust and reliable Web Services composition also represents one of the ten factors.

5   Conclusion

The intellectual structure of KM had been studied earlier by researchers in the Information Systems (IS) field. The finding of IS researchers is idiosyncratically inclined toward IS related research. This bias is probably due to their seminal author selection procedure, which is further compounded by the citation compilation process.

Our study draws on the CiteSeer citation index, which is primarily located in the fields of computer, information science, and engineering. The intellectual structure of knowledge management derived from a predominantly science and engineering oriented index is quite different from what has been provided by IS researchers. Our results reveal seventeen sub areas that form the conceptual groundwork of KM.

The current research trends of KM were briefly summarized. Research that intertwines World Wide Web and XML with classical AI topics seems to be the new direction. However, we have only seen limited new research that tries to leverage the rich AI tradition of the past to pursue Web related fields. Trust and security related issues are getting more attention due to the burgeoning Electronic Commerce and Electronic Business.

KM encompasses a fairly wide range of studies. It is also a field with great potential since knowledge has become the most important ingredient in modern businesses. The KM related research within science and engineering could provide the theoretical and infrastructural support that is needed by practitioners and researchers in other fields.

References

1. Earl, M.: Knowledge Management Strategies: Toward a Taxonomy. Journal of Management Information Systems 18 (2001) 215-242
2. Ponzi, L.J.: The Intellectual Structure and Interdisciplinary Breadth of Knowledge Management: a Bibliometric Study of Its Early Stage of Development. Scientometrics, Vol. 55 (2002) 259-272
3. Subramani, M., Nerur, S.P., Mahapatra, R.: Examining the Intellectual Structure of Knowledge Management, 1990-2002 – An Author Co-citation Analysis. Management Information Systems Research Center, Carlson School of Management, University of Minnesota (2003) 23
4. White, H.D., Griffith, B.C.: Author Cocitation: A Literature Measure of Intellectual Structure. Journal of the American Society for Information Science 32 (1981) 163-171
5. Chen, C.: Visualization of Knowledge Structures. In: Chang, S.K. (ed.): HANDBOOK OF SOFTWARE ENGINEERING AND KNOWLEDGE ENGINEERING, Vol. 2. World Scientific Publishing Co., River Edge, NJ, (2002) 700
6. Chen, C.: Searching for Intellectual Turning Points: Progressive Knowledge Domain Visualization. PNAS 101 (2004) 5303-5310
7. Chen, C., Kuljis, J., Paul, R.J.: Visualizing Latent Domain knowledge. Systems, Man and Cybernetics, Part C, IEEE Transactions on 31 (2001) 518-529
8. Chen, C., Paul, R.J.: Visualizing a Knowledge Domain's Intellectual Structure. Computer 34 (2001) 65-71
9. Culnan, M.J.: The Intellectual Development of Management Information Systems, 1972-1982: A Co-Citation Analysis. Management Science 32 (1986) 156-172
10. Culnan, M.J.: Mapping the Intellectual Structure of MIS, 1980-1985: A Co-Citation Analysis. MIS Quarterly 11 (1987) 340
11. McCain, K.W.: Mapping Authors in Intellectual Space: A Technical Overview. Journal of the American Society for Information Science 41 (1990) 433-443
12. White, H.D.: Pathfinder Networks and Author Cocitation Analysis: A Remapping of Paradigmatic Information Scientists. Journal of the American Society for Information Science & Technology 54 (2003) 423-434

13. Chen, C., Steven, M.: Visualizing Evolving Networks: Minimum Spanning Trees versus Pathfinder Networks. IEEE Symposium on Information Visualization. IEEE Computer Society (2003) 67-74
14. Bollacker, K.D., Lawrence, S., Giles, C.L.: CiteSeer: an Autonomous Web Agent for Automatic Retrieval and Identification of Interesting Publications. Proceedings of the second international conference on Autonomous agents. ACM Press, Minneapolis, Minnesota, United States (1998)
15. Chen, T.T., Xie, L.Q.: Identifying Critical Focuses in Research Domains. Proceedings of the Information Visualisation, Ninth International Conference on (IV'05). IEEE Computer Society, London (2005) 135-142
16. Ledru, P.: JSpace: Implementation of a Linda System in Java. SIGPLAN Not. 33 (1998) 48-50
17. White, H.D.: Author Cocitation Analysis and Pearson's r. Journal of the American Society for Information Science & Technology 54 (2003) 1250-1259

Evaluation of the FastFIX
Prototype 5Cs CARD System

Megan Vazey and Debbie Richards

Department of Computing, Division of Information and Communication Sciences,
Macquarie University
meganv@excelan.com.au, richards@ics.mq.edu.au

Abstract. The 5Cs architecture offers a hybrid Case And Rule-Driven (CARD) system that supports the Collaborative generation and refinement of a relational structure of Cases, ConditionNodes, Classifications, and Conclusions (hence 5Cs). It stretches the Multiple Classification Ripple Down Rules (MCRDR) algorithm and data structure to encompass collaborative classification, classification merging, and classification re-use. As well, it offers a very lightweight collaborative indexing tool that can act as an information broker to knowledge resources across an organisation's Intranet or across the broader Internet, and it supports the coexistence of multiple truths in the knowledge base. This paper reports the results of the software trial of the FastFIX prototype - an early implementation of the 5Cs model, in a 24x7 high-volume ICT support centre.

Keywords: Single Classification Ripple Down Rules, Multiple Classification Ripple Down Rules, SCRDR, MCRDR, Knowledge Engineering, Knowledge Acquisition, CARD, top-down rule-driven, bottom-up case-driven.

1 Introduction

The 5Cs model is comprised of the Collaborative generation and refinement of a relational structure of Cases, ConditionNodes, Classifications, and Conclusions (hence 5Cs). The 5Cs model is a Case And Rule Driven (CARD) system for Knowledge Acquisition (KA) that uses a Case Oriented Rule Acquisition Language (CORAL) to acquire rules in a similar way to Multiple Classification Ripple Down Rules (MCRDR) [7, 8] and its predecessor the Single Classification Ripple Down Rules (SCRDR) [4], but with significant extensions. For example, new data structures and algorithms are presented to allow experts to more effectively collaborate in building up both the knowledge and case bases.

The extensions have been motivated by our work in developing a trouble shooting system for a high volume ICT support centre. Knowledge in this domain changes rapidly and is driven by the need to maintain and link problem and solution cases. For this reason we chose to use the MCRDR combined case and rule based approach to incremental KA. However, in this domain we found that the knowledge needed to identify and solve the problem came from many, varied and globally distributed sources, and that the cases themselves were often in a state of flux and needed to be

A. Hoffmann et al. (Eds.): PKAW 2006, LNAI 4303, pp. 108–119, 2006.

worked up as new information came to light. Sources of information could include: clients, peers, third party vendors or software/hardware engineers and was obtained through personal conversations and notes, Internet sites, manuals and specifications.

The 5Cs system allows knowledge workers to collaboratively refine and expand a topic using an expert systems approach by consistently asking users to confirm, add to, or refine the knowledge presented, typically within the context of a current case. 5Cs supports intuitive KA as it allows the capture and sharing of the questions that experts ask themselves when classifying incoming cases.

This work is similarly motivated to the early work to reconcile multiple sources of expertise that have been captured into multiple MCRDR KBS [9] using Formal Concept Analysis [13]. In that work, however, identification and resolution of conflict was not fully integrated into the KA cycle, but performed as desired when another expert view was to be compared and required making changes to the individual sources and regeneration of the combined model for further verification and validation.

More recently Beydoun et al. [3] have extended Nested RDR (NRDR) [2] to support cooperative KA. NRDR differs from MCRDR in that it allows multiple SCRDR trees to be combined into an hierarchical conceptual structure. The approach seeks to detect *internal inconsistencies*, which are inconsistencies between multiple experts, and are resolved by the domain expert as independent KBS are individually integrated into the system. To assist, various estimates are derived to determine the probability of internal inconsistency and then determine a "trend of external inconsistencies". The degree of external inconsistency is seen to reflect how well the model represents the world in terms of ontological consistency, completeness and accuracy. As with the work described in this paper, "the intuition is that a coherent collective expertise is a better reflection of 'reality'" [3, p. 48]. Our work differs from this research in that combining expertise is a normal part of each KA cycle to develop a shared and single model, which can contain managed inconsistencies, rather than a technique employed to deal with inconsistencies whenever a new KBS is to be integrated. This is necessary in the 24x7 follow-the-sun call centre environment as multiple individuals, often distributed by time and place, may be involved with specifying the same problem/solution and need a way of working with the knowledge entered by themselves and others over time.

A subset of the features proposed by the 5Cs model has been tested via a prototype system known as FastFIX. FastFIX was developed to support the troubleshooting process in a high-volume and complex 24x7 support center in the ICT domain [11, 12]. The 5Cs system architecture is presented in the next section. The results of the FastFIX software trial are then evaluated and presented.

2 The 5Cs System

In the 5Cs system, a condition mesh structure is provided, comprising of RuleNodes i.e. (ConditionNodes) as shown in Fig 1. In this structure, each parent RuleNode may have multiple child RuleNodes, and each child RuleNode may have multiple parents. As well, there is an N-to-N relationship between Cases and their live and/or registered ConditionNodes (i.e. RuleNodes); an N-to-N relationship between the

ConditionNodes and the Classifications that they represent; and an N-to-N relationship between the resultant Classifications and their Conclusions.

A Case can evaluate to TRUE for multiple Condition paths in the condition mesh, hence a Case can fetch multiple Classifications, where each Classification may be linked to multiple Conclusions. As well, Conclusions can be reused across multiple Classifications, and those Classifications may be reused across Condition paths, and across multiple Cases. Note that in Fig 1 only a subset of classifications and conclusions are shown for RuleNodes 6, 7, and 8 and for simplicity links to the classifications and conclusions at other RuleNodes have not been included in the figure. Note also that in the 5Cs system, the Attributes, Cases, Conditions, Classifications, and Conclusions may each be the subject of Collaborative creation, editing, or deletion. In addition, RuleNodes can be collaboratively relocated.

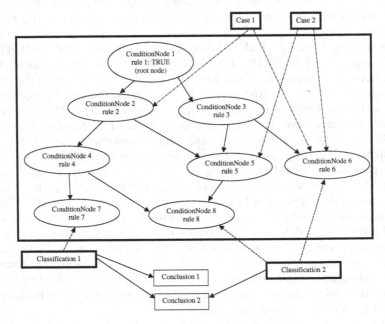

Fig. 1. 5Cs Condition Mesh

2.1 Tracking Case-RuleNode Associations

Unlike many MCRDR systems which only require cornerstone cases to be kept, in a domain like the call center where the cases are volatile we keep all cases and track all Case-RuleNode associations. This is somewhat similar to the use of an execution history, tracking of rule usage and the proposed review of rule usage against the case history in the HeurEAKA RDR system [1], which uses NRDR and genetic algorithms for channel routing in VLSI design. In the 5Cs model, a "Tracked" Case is one whose Live and Registered RuleNodes are being remembered by the system. A Live RuleNode for a given Case is one that is currently the last TRUE RuleNode on a given path the knowledge base for that case. Its conclusions are part of the set of current conclusions derived from the knowledge base for the case. The system remembers its Live RuleNodes for "Tracked" cases. Live RuleNodes may be correct or incorrect. Correcting incorrect live RuleNodes is the primary role of a (human or computer) expert who trains the knowledge base.

A Registered RuleNode is one that has been confirmed by a User as being correct and TRUE for that Case. For each case, each RuleNode registration may be current or expired. The test for RuleNode-Case registration expiry is that if the last modification or creation date on the RuleNode-Case association is more recent than or the same as the last modified dates on the RuleNode and the Case, then the registration is current. Otherwise, the registration is expired. The expiry of registered case-RuleNode relationships is something that can be notified and displayed to users in the summary of cases or RuleNodes of interest to them. The user can also be notified whenever the list of live RuleNodes differs from the list of registered RuleNodes for a given case, or the list of live cases differs from the list of registered cases for a given RuleNode.

In the Pathology Expert Interpretative Reporting System (PEIRS), typographical or conceptual errors in RuleNode expressions were corrected with the use of "fall-through" rules [5, p. 119], but potentially resulting in corruption of the KBS [5, p.88]. Similarly, Kang identifies the situation where the domain knowledge represented by an existing rule tree needs to be changed in such a fashion that a cornerstone case for an existing RuleNode will *drop-down* to a new child RuleNode [7, p.50]. He suggests that if absolutely necessary, the rules suggested by the MCRDR difference list for the new RuleNode can be overridden [7, p. 65]. The approach was not fully explored [7, p. 67]. Note that cases don't only *drop-down*. They may drop-across from a sibling node; or they can *recoil* to an ancestor RuleNode for example when the editing or relocation of a dependent RuleNode is restrictive enough that the case under review is now excluded.

Unlike in pathology where each report handled a new case and changes to a patients data resulted in a new case with new conclusion, in the Call Centre domain it is likely that information about a customer problem will develop over days or even months in situations where a problem is intermittent or temporarily fixed but reemerges at a later date. A supposed solution to a problem may turn out to not have really fixed the problem at all. This means that the knowledge (ie the rules) and the problem/solution case/s may also need changing. The 5Cs structure and algorithms allow the knowledge base to evolve, including changes to cases, RuleNodes, intermediate conclusions, ontological entry or attributes. This is achieved through tracking of live vs registered case-RuleNode associations.

In fact, FastFIX tracks all changes to the system and identifies users when an inconsistency in the system occurs. The strategy assists with more rapid knowledge acquisition since it can highlight inconsistencies between expert opinions, just as MCRDR supported quicker KA by allowing more than one rule to be acquired for each case. It lets users capitalize on the knowledge acquisition opportunity presented by the case drop-down scenario, and this in turn may result in quicker coverage of the domain and greater learning opportunities for users. The separation of live and registered case-RuleNode associations is a key part of being able to resolve classification conflicts between multiple experts, and even between what a single expert thinks today, as compared with tomorrow [6].

2.2 The FastFIX Prototype

As mentioned earlier, a subset of the features proposed by the 5Cs model was tested in the ICT support center problem domain via a prototype system known as FastFIX. Significant novel ideas implemented in the FastFIX prototype and tested during the FastFIX software trial included:

- The ability for multiple users to build an MCRDR-based decision tree in a wiki[1]-style collaborative effort. This includes the identification of classes of incoming problem cases and manual indexing of solutions by multiple users using rule conditions equivalent to logical tags in a folksomony[2].
- Reference to multiple exemplar cornerstone cases for each RuleNode.
- The ability for users to edit previously created cases (including cornerstone cases) and RuleNodes in the system.
- Continuous background monitoring of changes to the knowledge base so that users with affected RuleNodes and Cases can notice and respond to the changes. This approach allows classification conflicts to be identified, clarified and resolved and hence it enhances knowledge acquisition.
- The ability for users to "work-up" a case using a novel Interactive and Recursive MCRDR decision structure.
- Separation between classifications and conclusions so that richer classification relationships can be maintained.
- The ability for users to relocate i.e. move RuleNodes in the system.

3   Results of the FastFIX Software Trial

Table 1 summarises user activity during the software trial. 12 users registered themselves, including the author (User ID 12). Most of the registered users were onlookers – managers, team leaders and other interested parties. In addition to the author, there were three main contributors with user IDs 1, 3 and 6. These contributors were able to use the system with minimal training and supervision (less than 60 minutes per contributor). Contributors commenced by providing troubleshooting knowledge for the most frequently occurring sub-domain of troubleshooting errors.

In total 172 cases and 107 RuleNodes were created. The total number of case edits was 139 and the total number of RuleNode edits was 141 demonstrating both the desire and capacity of users to contribute to knowledge evolution in this way. In total there were 104 case drop-throughs resulting from RuleNode creations. As well, there were 96 case drop-through events resulting from RuleNode edits where each of these events may have affected 1 or more cases. Fig. 2 provides a graphical representation of the data in Table 1.

---

[1] Wikipedia (http://www.wikipedia.org/) defines a Wiki as the collaborative software and resultant web forum that allows users to add content to a website and in addition, to collaboratively edit it. Wikipedia demonstrates the power of the Wiki paradigm.

[2] The term "folksomony"[2] was first coined by Thomas Vander Wal[2] (2005) to describe forums in which people can tag anything that is Internet addressable using their own vocabulary so that it is easy for them to re-find the item.

Table 1. User Activity in the prototype FastFIX system

User ID	1	2	3	4	5	6	7	8	9	10	11	12	Totals
Total Case Creations	64	0	15	2	2	83	1	1	0	0	1	3	172
Total RuleNode Creations	42	0	13	0	0	30	1	0	0	0	0	21	107
Total Case Edits	45	0	22	0	0	33	0	0	2	0	7	30	139
Total RuleNode Edits	32	0	13	0	0	59	0	0	0	0	0	37	141
Total Case drop-throughs resulting from RuleNode Creations	59	0	1	0	0	43	0	0	0	0	0	1	104
Total Case drop-through Events resulting from RuleNodeEdits	24	0	6	0	0	48	0	0	0	0	0	18	96

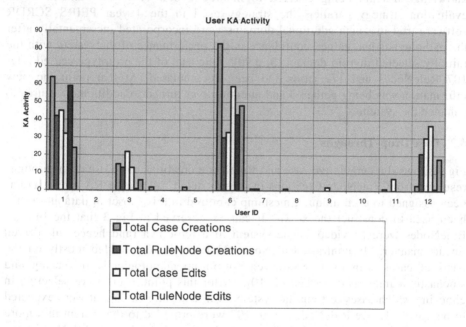

Fig. 2. User Activity in the prototype FastFIX system

3.1 Solution Effectiveness

After 7 hours of effort the test team had captured 105 cases and 55 RuleNodes. The red arrow in each of following figures has been used to indicate this point in time. At this point the team had provided enough RuleNodes to automatically solve approximately 90% of errors on errant equipment in the selected error sub-domain. These errors contribute to 30% of all errors seen by the system which account for 20% of the ~5,000 problem cases per day seen by the global ICT support centre. Hence after 7 hours of effort enough knowledge had been acquired to automatically provide solutions to more than 270 cases per day, without requiring the trouble-shooters to figure out the class of problem on hand, where to search for a solution, or what to search for, for example in the corporate solution tracking system.

Say that each case takes on average 15 minutes to solve, and that 1 minute of this time is spent in determining the problem and finding its solution. This represents a time saving of 1 mins * 270 cases per day, or 4.5 hours per day. Actually, the average solution search time is possibly a lot longer. One of the problems with manually searching for solutions is that if you haven't found the answer, you don't know if its just because your not searching for it correctly, or if its because a solution does not exist. The FastFIX system has the advantage that it unambiguously associates relevant solutions with their incoming problem classes. If the answer is unknown, FastFIX can provide that information.

After the first 105 cases and 55 RuleNodes, the test team broadened the knowledge domain being covered to include a new error sub-domain. This evaluation strategy parallels the strategy used in the 4 year PEIRS SCRDR software trial in which additional domains were incorporated incrementally after the pathologists had gained confidence in the performance of the system with the initially selected thyroid domain [5, p.90]. The trial of the prototype ceased after 107 RuleNodes and 172 cases had been accumulated. At that point, no new information was being gathered and attention was turned to additional features to enhance the system.

3.2 Case Drop-Throughs

Fig 3 shows the cumulative case and RuleNode creations and Case drop-throughs resulting from RuleNode Creations in greater detail. A unique KA Event ID has been assigned to each unique timestamp captured in this subset of data and it has been used to construct the x-axis. It can be observed in Fig 3 that the first 20 RuleNodes were provided to the system in a top-down (and hence rule-driven linear) manner. In contrast, RuleNodes 21 to 55 were provided mostly on the basis of cases seen in a case-driven bottom-up monotonically increasing and stochastic manner as described in [10]. After this point, users were selective in choosing which cases to train the system with, choosing cases that were expected to be novel. Hence RuleNodes 56 to 107 were provided to the system in a more top-down (and hence rule-driven linear) manner as for the first 20 RuleNodes.

It is difficult to say how the ability of users to edit RuleNodes affects the overall case and RuleNode creation trajectories. If most of the RuleNode edits were cosmetic e.g. as a result fixing spelling mistakes then it can be expected that these KA trajectories would be little affected by the RuleNode edits. However, if RuleNode editing represents a significant KA activity whereby genuinely new knowledge is being acquired, rather than existing knowledge being cosmetically corrected, then those RuleNode edit events should be added into the above case-driven KA trajectory. However, it was beyond the scope of this trial to examine this in any detail.

Fig 4 shows the Cumulative Case Creation and RuleNode Edit Curves. The number of RuleNode edits appears to grow in proportion to the number of cases seen by the system, which indicates that RuleNode editing tends to be a top-down knowledge acquisition activity.

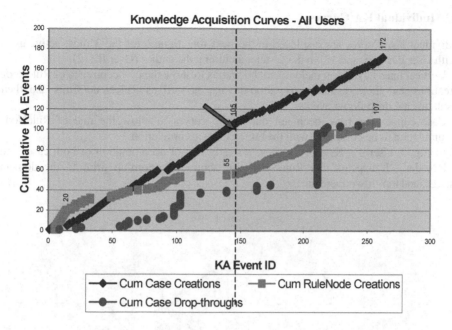

Fig. 3. Cumulative Case and RuleNode Creation Curves

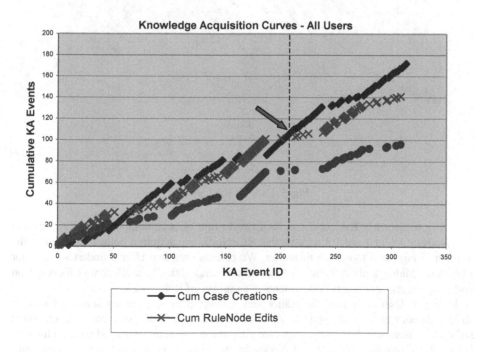

Fig. 4. Cumulative Case Creation and RuleNode Edit Curves

3.3 Individual KA Curves

Individual KA Curves are displayed in the next four figures for the 3 most active users (with User IDs 1, 6, and 12) in the system, including the author (User ID 12).

Vertical lines have been included in the graphs to show the co-occurrence of RuleNode Creations and their resultant case drop-downs, as well as RuleNode Edits and their resultant case drop-down events.

Case edit events have been left out of the curves to allow the rate of RuleNode accumulation to be compared with the rate of Case accumulation.

In Fig. 5, User 1's KA curves show a steady rate of accumulation of both cases and RuleNodes. It appears that RuleNodes are acquired bottom-up prior to the domain change, and top-down thereafter.

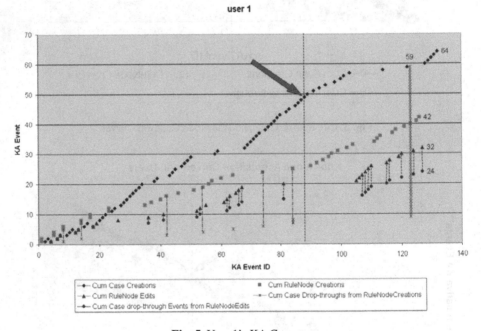

Fig. 5. User 1's KA Curves

In Fig 6, User 6's KA curves show a steady rate of accumulation of both cases and RuleNodes. As for User 1 it appears that RuleNodes are acquired bottom-up prior to the domain change, and top-down thereafter. We can also see that User 6 undertook a major RuleNode editing activity between KA event 80 and 100. This effort was focussed on widening the scope of the rule statements in a number of RuleNodes.

In Fig 7, User 12's (i.e. the author and researcher's) KA curves show a focus on RuleNode edits in the early phases. At this point the system was still under development so both the users and the system were changing in the way they interacted with each-other. User 12 created the first 20 RuleNodes in the system in a top-down fashion after consulting with User 6. In contrast, User 12 was involved in very few case creations.

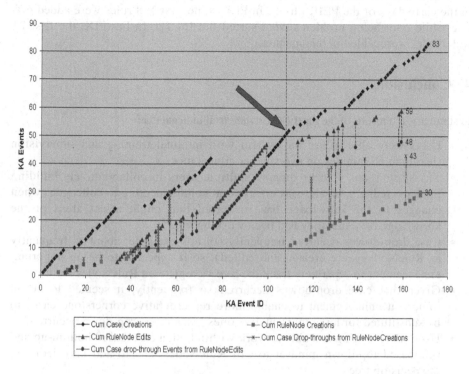

Fig. 6. User 6's KA Curves

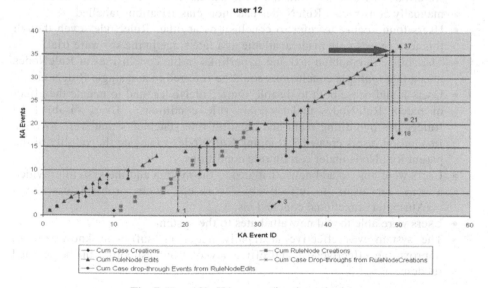

Fig. 7. User 12's KA curves (i.e. the author's)

Note that the initial knowledge base activity by user 12 parallels that reported in the early days of the PEIRS trial. In PEIRS the first 198 rules were added off-line while interfacing problems were sorted out [5, 7]. In FastFIX, the first 21 RuleNodes were added in this manner.

4 Conclusions

In summary, the results of the FastFIX software trial indicate that:

- Users were able to use the system with minimal training and supervision (less than 60 minutes on average per contributor).
- The system was able to support multiple users in collaboratively building the knowledge base, and was effective at communicating to colleagues when changes to the knowledge base occurred that might affect areas of the knowledge base that they had been working on.
- Case drop-downs occurred frequently (in this example, about as frequently as RuleNodes were created and edited), so it appeared to be an important enhancement to separately track the live vs registered RuleNodes for users.
- Given that case drop-downs occurred so frequently, it seemed to be an important enhancement to enable more representative cornerstone cases to be substituted for less representative ones when case drop-down occurred.
- Users were willing and able to work within both a case-based bottom-up and rule-based top-down mindset to edit RuleNodes, and to add knowledge to the decision tree.
- Users took advantage of the ability to label the classification represented by RuleNodes as distinct from the conclusions at that RuleNode. 22 of 106 manually constructed RuleNodes had their classifications labelled.
- Users found cause to refer to conclusions at other RuleNodes even though this feature was only made available at a late stage in the software trial.
- The ability to combine text and hyperlinks in the conclusions at RuleNodes meant that multiple conclusions could be referred to at a single RuleNode.
- Users found cause both to disable some RuleNodes, and to negate the effect of parent RuleNodes under certain rule conditions. Users disabled 10 RuleNodes by editing them and creating rule statements that were FALSE. In addition, users created 3 stopping RuleNodes to negate the validity of the parent RuleNode under certain rule conditions.
- Users were able to add new attributes to the system and the system provided a way for users to effectively "work-up" their problem cases via getAttribute() functional conclusions.
- Users were able to add new attributes to the system.
- The system was effective at rapidly acquiring sufficient knowledge to improve the effectiveness and efficiency of troubleshooting in the selected subdomains.

Acknowledgments. This work has been funded via an Australian Research Council Linkage Grant LP0347995 and a Macquarie University RAACE scholarship. Thanks also to my industry sponsor for ongoing financial support.

References

1. Bekmann, J. and Hoffmann, A. (2004) HeurEAKA– A new approach for Adapting GAs to the problem domain, Eds. Zhang, Guesgen, Yeap *PRICAI 2004: Trends in Artificial Intelligence*, Springer, Berlin, 2004, pp. 361 - 372 .
2. Beydoun, G. and Hoffmann, A. (2000). Incremental acquisition of search knowledge. *Journal of Human-Computer Studies*, 52:493.530, 2000.
3. Beydoun, G., Hoffmann, Fernandez Breis, J. T, Martinez Bejar, R. Valencia-Garcia, R. and Aurum, A. (2005) Cooperative Modeling Evaluated, *International Journal of Cooperative Information Systems*, World Scientific, 2005, 14 (1), 45-71.
4. Compton, P. J. and R. Jansen (1989). A philosophical basis for knowledge acquisition. 3rd European Knowledge Acquisition for Knowledge-Based Systems Workshop, Paris: 75-89.
5. Edwards, G. (1996) Reflective Expert Systems in Clinical Pathology MD Thesis, School of Pathology, University of New South Wales.
6. Gaines B. R., Shaw M. L. G., (1989) Comparing Conceptual Structures: Consensus, Conflict, Correspondence and Contrast.
7. Kang, B. (1996) Validating Knowledge Acquisition: Multiple Classification Ripple Down Rules PhD Thesis, School of Computer Science and Engineering, University of NSW, Australia.
8. Kang, B., P. Compton and P. Preston (1995). Multiple Classification Ripple Down Rules : Evaluation and Possibilities. Proceedings of the 9th AAAI-Sponsored Banff Knowledge Acquisition for Knowledge-Based Systems Workshop, Banff, Canada, University of Calgary.
9. Richards, D and Menzies, T. (1998) Extending the SISYPHUS III Experiment from a Knowledge Engineering to a Requirements Engineering Task, Richards D., Menzies, T., Task 11th Workshop on Knowledge Acquisition, Modeling and Management, SRDG Publications, Banff, Canada, 18th-23rd April, 1998.
10. Vazey, M. (2006) Stochastic Foundations for the Case-driven Acquisition of Classification Rules. *EKAW 2006* (in publication).
11. Vazey, M. and Richards, D. (2004) Achieving Rapid Knowledge Acquisition. Proceedings of the Pacific Knowledge Acquisition Workshop (PKAW 2004), in conjunction with The Eighth Pacific Rim International Conference on Artificial Intelligence, August 9-13, 2004, Auckland, New Zealand, 74-86.
12. Vazey, M. and Richards, D. (2006) A Case-Classification-Conclusion 3Cs Approach to Knowledge Acquisition - *Applying a Classification Logic Wiki to the Problem Solving Process*. International Journal of Knowledge Management (IJKM), Vol. 2, Issue 1, pp 72-88; article #ITJ3096, Jan-Mar 2006.
13. Wille, R. (1992) Concept Lattices and Conceptual Knowledge Systems *Computers Math. Applic.* (23) 6-9: 493-515.

Intelligent Decision Support for Medication Review

Ivan Bindoff[1], Peter Tenni[2], Byeong Ho Kang[1], and Gregory Peterson[2]

[1] University of Tasmania, School of Computing
{ibindoff, bhkang}@utas.edu.au
[2] University of Tasmania, Unit for Medical Outcomes and Research Evaluations
{Peter.Tenni, G.Peterson}@utas.edu.au

Abstract. This paper examines an implementation of a Multiple Classification Ripple Down Rules system which can be used to provide quality Decision Support Services to pharmacists practicing medication reviews (MRs), particularly for high risk patients. The system was trained on 84 genuine cases by an expert in the field; over the course of 15 hours the system had learned 197 rules and was considered to encompass around 60% of the domain. Furthermore, the system was found able to improve the quality and consistency of the medication review reports produced, as it was shown that there was a high incidence of missed classifications under normal conditions, which were repaired by the system automatically.

1 Introduction

Sub-optimal drug usage is a serious concern both in Australia and overseas [1, 2], resulting in at least 80,000 hospital admissions annually - approximately 12% of all medical admissions and reflecting a cost of about $400 million annually, with the majority of these affecting elderly patients [3]. MRs are seen as an effective way to improve drug usage. However, the quality of MRs produced is inconsistent across reviewers. Further to this, many community-based pharmacists are still unwilling to undertake this new role, citing reasons including fear of error and a lack of confidence [4].

This paper proposes a different approach to improving the quality of the MRs, and possibly even improving the uptake of the role within the pharmaceutical community. It is suggested that the answer may lie in the development of medication management software which includes Intelligent Decision Support features. To date, the majority of incarnations of medication management software for producing MRs has lacked any form of genuine Decision Support features [5]. Unfortunately, Knowledge Based System (KBS) techniques which may be suitable to this problem have been designed to handle steadfast, well defined sets of knowledge, and have historically not been well suited to poorly structured or dynamic sets of knowledge such as the set found in the domain of MR. However, newer techniques such as Case Based Reasoning (CBR) and Ripple Down Rules (RDR) may offer new possibilities in handling knowledge of this kind, since they are easily, even naturally, maintainable and alterable [6, 7].

A. Hoffmann et al. (Eds.): PKAW 2006, LNAI 4303, pp. 120–131, 2006.
© Springer-Verlag Berlin Heidelberg 2006

2 Medication Reviews

MR is a burgeoning area in Australia and other countries, with MRs seen to be an effective way of improving drug usage and reducing drug related hospital admissions, particularly in the elderly and other high risk patients [1, 3]. This has prompted the Australian government to initiate the Home Medicines Review scheme (HMR) and the Residential Medication Management Reviews (RMMRs) scheme. These schemes provide remuneration to pharmacists performing MRs via a nationally funded program [3]. However, it is known that despite Residential Medication Management Reviews (RMMRs) being introduced in 1997 they still do not have a conceptual model for delivery, which has resulted in a wide range of differing qualities of service being provided [4].

To perform a MR, Pharmacists assess potential Drug Related Problems (DRPs) and Adverse Drug Events[1] (ADEs) in a patient by examining various patient records, primarily their medical history, any available pathology results, and their drug regime (past and current) [8]. The expert looks for a variety of indicators between the case details provided checking for known problems, such as an: Untreated Indication – where a patient has a medical condition which requires treatment but doesn't have the treatment; Contributing Drugs – where a patient has a condition and is on a drug which can cause or exacerbate said condition; High Dosage – where a patient is potentially on a too high dosage because of a combination of drugs with similar ingredients; Inappropriate Drug – where a patient is on a drug that is designed to treat a condition they don't seem to have or is contraindicated in their condition; and many others besides. Once these indicators have been identified a statement is produced explaining each problem, or potential problem, and often what the appropriate course of action is.

3 Methodology

In order to produce a medication management system with intelligent decision support features it was necessary to produce two major software elements. The first was a standard implementation of a database "front-end" from which it is possible for a user to enter all the details of a given patient's case, or at least those parts which are relevant to the chosen domain. The second was an implementation of a Multiple Classification Ripple Down Rules engine which can sufficiently encapsulate the types of conditions and knowledge required for the domain and facilitate the design of an interface from which the engine can be operated, particularly during the Knowledge Acquisition phase.

3.1 Database

The design of the database to store the MR cases was relatively trivial, and will not be given much detail here. The preliminary design idea was taken from existing

---

[1] Defined by the World Health Organisation as being "an injury resulting from medical intervention related to a drug." 2. Bates, D., et al., *Incidence of adverse drug events and potential adverse drug events. Implications for prevention. ADE Prevention Study Group.* JAMA, 1995(274): p. 29-34.

medication management software packages, and then extensively modified to allow for proper computerized analysis. The 126 cases considered in this study were then inserted into the database using a simple script which converted them from their current Mediflags [9] format.

3.2 Ripple Down Rules

Ripple Down Rules (RDR) is an approach to building KBSs that allows the user to incrementally build the knowledge base while the system is in use, with no outside assistance or training from a knowledge engineer [7]. It generally follows a forward-chaining rule-based approach to building a KBS. However, it differs from standard rule based systems since new rules are added in the context in which they are suggested.

Observations from attempts at expert system maintenance lead to the realisation that the expert often provides justification for why their conclusion is correct, rather than providing the reasoning process they undertook to reach this conclusion. That is, they say 'why' a conclusion is right, rather than 'how'. An example of this would be the expert stating "I know case A has conclusion X because they exhibit features 1, 4 and 7". Furthermore, experts are seen to be particularly good at providing comparison between two cases and distinguishing the features which are relevant to their different classifications [10]. With these observations in mind an attempt was made at producing a system which mimicked this approach to reasoning, with RDR being the end result.

3.3 Structure

The resultant RDR structure is that of a binary tree or a decision list [11], with exceptions for rules which are further decision lists. The decision list model is more intuitive since, in practice, the tree would have a fairly shallow depth of correction [12]. The inferencing process works by evaluating each rule in the first list in turn until a rule is satisfied, then evaluating each rule of the decision list returned by that satisfied rule similarly until no further rules are satisfied. The classification that was bound to the last rule that was satisfied is given.

RDR can be viewed as an enhancement to CBR [6, 13, 14], with RDR providing a utility, in the form of an algorithm, a structure and rules, with which to demonstrate which parts of the case are significant to a particular classification [15].

3.4 Multiple Classification Ripple Down Rules

The RDR method described above is limited by its inability to produce multiple conclusions for a case. To allow for this capability - as this domain must - MCRDR should be considered [16] to avoid the exponential growth of the knowledge base that would result were compound classifications to be used.

MCRDR is extremely similar to RDR, preserving the advantages and essential strategy of RDR, but augmented with the power to return multiple classifications. Contrasting with RDR, MCRDR evaluates all rules in the first level of the knowledge base then evaluates the next level for all rules that were satisfied and so on, maintaining a list of classifications that should fire, until there are no more children to evaluate or none of the rules can be satisfied by the current case [12]. An example of this can be seen in Fig. 1.

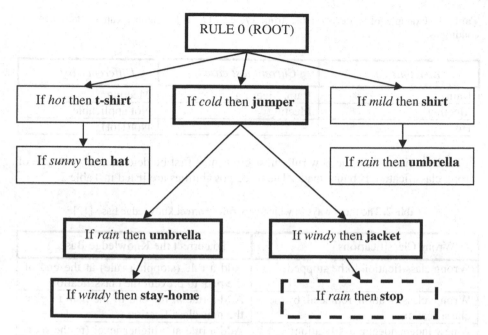

Fig. 1. The highlighted boxes represent rules that are satisfied for the case (cold, rain, windy), the dashed box is a potential stopping rule the expert may wish to add [17]

3.5 Knowledge Acquisition

Knowledge Acquisition is required when a case has been classified incorrectly or is missing a classification. It is divided into three separate steps: Acquiring New Classi-fication (or Conclusion), Locating the New Rule, and Acquiring the New Rule. It should be noted that the order of applying steps one and two is unimportant to the validity of the method [12], and that for the purposes of this experiment it made sense to locate the rule before acquiring the new classification. Hence this is what was done.

Acquiring the New Classification is trivial; the system merely prompts the expert to state it [12]. To Acquire the New Rules the expert is asked to first select valid con-ditions from the current case that indicate a given classification. The rule they have created thus far is then compared against the cornerstone case base. If any cornerstone cases would fire on this new rule the expert is asked to select from a difference list (see Table 1) between the presented case and one of the cornerstone cases. A corner-stone case is a case for which the knowledge had previously been modified and which is valid under the current context [18]. The system then re-tests all cornerstone cases in the list against the appended set of conditions, removing cases from the list that are no longer satisfied. The system repeats this process until there are no remaining cor-nerstone cases in the list to satisfy the rule [12] or alternatively the expert has stated explicitly that the cornerstone cases that remain *should* fire on the new rule and this new classification was simply missed when the cornerstone case was originally considered.

Table 1. Example of a decision list from [7, 15, 17, 19]. The list can contain negated conditions.

Cornerstone case	Current test case	Difference list
Rain	Rain, Meeting	Meeting
Meeting	Meeting	Not applicable
Hot		Not(Hot)

To determine where the new rule must go it must first be determined what type of wrong classification is being made. The three possibilities are listed in Table 2.

Table 2. The three ways in which new rules correct knowledge base [12]

Wrong Classifications	To correct the Knowledge Base
Wrong classification to be stopped	Add a rule (stopping rule) at the end of the path to prevent the classification
Wrong classification replaced by new classification	Add a rule at the end of the path to give the new classification
A new independent classification	Add a rule at a higher level (to the root) to give the new classification

4   Results and Discussion

The system was handed over to the expert with absolutely no knowledge or conclusions pre-loaded. The expert was wholly responsible for populating the knowledge base. Over the course of approximately 15 hours they were able to add the rules required to correctly classify 84 genuine MR cases that had been pre-loaded into the system.

4.1   Growth of Knowledge Base

It is observed in Fig. 2 that the number of rules in the system progressed linearly as more cases were analysed, at a reasonably consistent rate of about 2.3 rules per case. This suggests that the system was still in a heavy learning phase when the experiment was finished, since it has previously been observed that RDR systems will show a flattening pattern in the rate of growth of the knowledge base at approximately 80% of domain coverage [12]. This has complications for many of the remaining tests, in that their results must be understood to reflect the knowledge base while it is still learning heavily. The general conclusion that can be applied here is that most results will be expected to improve with additional testing, and that further testing is indeed required. This is because without demonstrating that the rate of learning has begun to slow down it is impossible to adequately prove that the heavy learning phase, which requires a significantly higher level of expert maintenance, will cease.

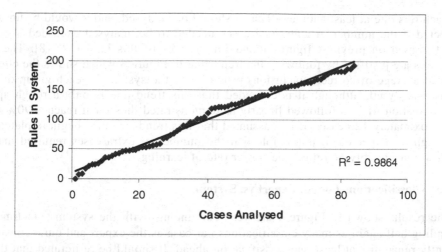

Fig. 2. The number of rules in the system grows linearly as more cases are analysed

4.2 Correct Conclusions Found

It was estimated by the expert at the time of cessation of the experiment that the system had encapsulated around 60% of the domain [20], this estimation is supported by the evidence shown in Figure 3. It can be seen that the average number of correct classifications the system provided rose quite steadily into the 60th percentile, although the percentage correct from case to case did vary quite a lot, as is to be expected when the system is still in the heavy learning phase.

The expert predicted potential classification rates in the order of 90% [20], so considering 84 cases had been analysed it could be estimated that in order to

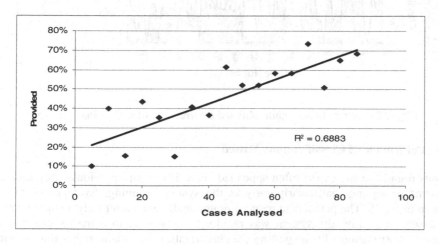

Fig. 3. The percentage of conclusions provided by the system for each case that were already correct

reach this rate at least another 40 cases should be analysed, and it would be unexpected if the number of additional cases needed to be analysed exceeded about 120, based on previous figures found for systems of this kind [12, 18]. These figures are justified by following the trend-line in Figure 3 which shows the clustered average of correct conclusions provided by the system for each group of 5 cases analysed, although it is conceded that this trend-line is only a rough approximation. If it is followed linearly as demonstrated thus far it reaches 90% at approximately 120 cases, if it is assumed that this trend-line may begin to plateau though, as expected, it is possible that the number of extra cases required may grow considerably, to reflect the slower rate of learning.

4.3 Classifications Found, Expert vs. System

The results shown in Figure 4 are very convincing, with the system sometimes finding half again as many classifications per case as the expert and quite consistently remaining at least one classification ahead. It should be re-iterated that the system found all these classifications using only a smaller set of the same knowledge the expert had. This suggests the expert consistently misses classifications they should find. In other words, they just don't notice them on the particular case. The system does not suffer from this, it will notice anything that it is trained to know about without exception.

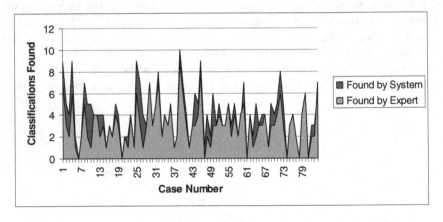

Fig. 4. The system found significantly more correct classifications than the expert

4.4 Percentage of Classifications Missed

It was found that the expert often appended classifications to previous cases after the systems prompting, particularly early in the systems training. Evidence of this is shown in Fig. 5. The percentage reduced dramatically even after only a small number of cases, suggesting the system was rapidly helping to reduce the experts rate of missed classifications, by suggesting the classifications for them, rather than making the expert notice themselves. The trend-line in Fig. 5 is only an approximation, since relatively few cases have been analysed thus far, and noise is still significant.

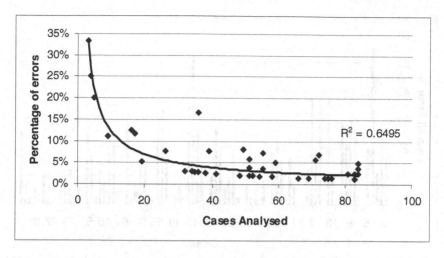

Fig. 5. The percentage of cases that gained new classifications

4.5 Total Errors per Case

It was found that the rate of error in each case was quite high, averaging 13.4% and with some going over 50%. Clearly the expert is making errors regularly, as was expected, and yet these numbers would be expected to be even higher were more complete training done. It is important to note that the results shown in Fig. 6 are representative of all the errors (missed classifications) that the system has fixed through the normal course of operation, and not the actual number of errors per case. This figure suggests that over 1 in 9 classifications are missed, although it is unclear what type of classifications these are. What level of threat are these classifications likely to pose? One would like to assume that the expert would not miss life threatening classifications, because they would have a particular focus on these, but additional experimentation is clearly required to determine what kinds of classifications the expert is missing and what the consequences of this is.

4.6 Maintainability and Usability

It was considered possible that the domain of MR might damage the maintainability and usability of the system due to both its inconsistent/dynamic nature, and the large number of variables within each case. It was considered possible that the dynamic nature of the domain might result in a need for an excessive number of exceptions to be added to the knowledge base, and that the large number of variables within each case may have resulted in an excessive number of conditions required for each rule. Each of these afflictions would increase the time taken to maintain/use the system and possibly make it untenable. As such, tests were carried out to determine whether these figures deviated remarkably from the normal range to be expected in a system of this nature.

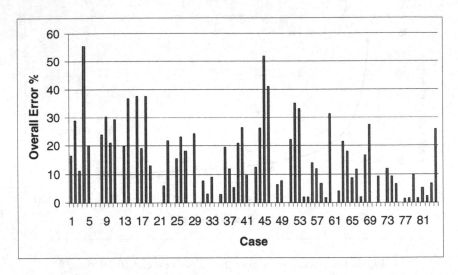

Fig. 6. The final percentage of classifications missed by expert per case

Time Spent: Adding Rules

Previous RDR systems have reported figures of around 3 minutes per rule [18], and it was found that this system continued that trend, with the average time taken to add a rule being 183 seconds (3 minutes).

Time Spent: Analysing Cases

It was found that the average time taken for the expert to complete a case analysis was about 10 minutes (621 seconds). This average extended over the entire 84 cases gives a total expert time taken as about 15 hours as reported earlier. Some cases were done in as few as 2 minutes when no or few new classifications were required, although the process did sometimes reach over 20 minutes.

Cornerstones Seen

The results here are promising from a usability point of view, with the expert rarely having to consider cornerstone cases in the creation of rules, with the majority of rules having no cornerstone cases to consider. In fact the expert saw an average of only 0.42 cornerstones per rule. What this means is that the expert should be able to add rules relatively quickly, with the time required to validate their rules being small.

4.7 Structure of the Knowledge Base

It can be determined from Table 3 that the structure of the knowledge base tree was extremely shallow and branchy, meaning the possibility of an excessive number of exceptions has not, at least at this point, come to light at all.

Table 3. Structure of the Knowledge Base Tree

Tree Property	Value
Average Depth	1.30
Depth 1	139
Depth 2	53
Depth 3	3

The nature of the rules in the knowledge base is also of interest, with further support for the maintainability of the system shown in the fact that the average number of conditions selected in a rule was only 1.7, with longer rules of 4 or 5 conditions being virtually non-existent and no rules with 6 or more being present.

To get a more complete view of the knowledge base it is necessary to analyse what outputs the rules map to. With the knowledge base that was built in the process of this experiment 85 individual conclusions were defined. When it is considered that every rule except stopping rules, of which there are 154, is linked to a conclusion it can be seen that there is 1 conclusion for every 1.8 rules, as can be demonstrated with the data used in Fig. 7. It is evident from this figure that, although most conclusions are only used by one rule, some conclusions are used very often. In other words they have many different sets of conditions which can lead to them.

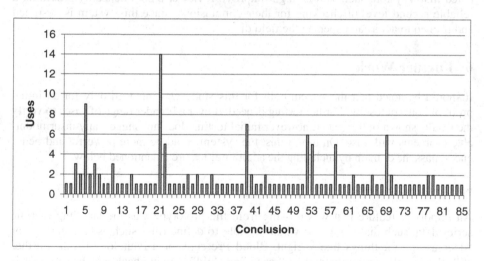

Fig. 7. How many times each conclusion was used

5 Conclusions

Initial experimentation suggests that the proposed method using MCRDR can successfully represent knowledge where the knowledge sources (human experts) are inconsistent. The system is shown to have reached about an 60% classification rate with less than 15 expert hours and only 84 cases classified – a good outcome in the circumstances. The knowledge base structure does not show any major deviations from what would be anticipated in a normal MCRDR system at this stage. The maintainability of the system does not appear to have been adversely affected thus far, with the expert being faced with only few cornerstone cases during the knowledge base validation, and the time taken to add rules being negligible.

From a MR perspective the system is seen to be capable of: providing classifications for a wide range of Drug Related Problems; learning a large portion of the domain of MRs quickly; producing classifications in a timely manner; and importantly, vastly reducing the amount of missed classifications that would otherwise be expected of the reviewer. It is expected that this system, or a future incarnation of this system, would be capable of achieving classification rates around 90% [20]. If this figure is to be realised it is possible that this system would be capable of achieving three major goals:

- Reducing the amount of missed classifications
    o Thus improving the consistency (quality) of service
- Improving the confidence of potential medication reviewers

It has already been noted that the number of errors this system detected and repaired was significant, and the number of errors was seen to reduce as the expert populated the knowledge base and this result alone would be enough to warrant further work. It has also been observed that the amount of time taken to perform a MR using this system should not be adversely affected. As for the final point it is anticipated that a system such as this might improve reviewer's confidence by providing a reliable second level of checking for their conclusions, since this system is designed and trained to act as an expert in the field did.

6 Further Work

It should be noted that the system built for this study was intended only for an initial proof of concept testing. Further testing is needed over a broader range of cases to verify the results shown in this paper, however initial testing does not suggest any insurmountable problems will arise. On top of this, the system could be more powerful and better encompass the domain by including the additional features mentioned below.

6.1 Time Series Data

An important feature that was missing from the prototype was the handling of time series data, such that the expert would be able to define rules such as "increasing" or "decreasing" for things like Weight, Blood Pressure, or a Pathology result. Further still, they might define things like "recent" or "old", which check whether a result is older or younger than defined thresholds, newest, oldest, average and others. As the system stands it will fire on a rule that states "Creatinine > 0.12" even if the result which says their Creatinine level was 0.13 was taken 15 years prior. This is undesirable, with the meaning of the results varying across periods of time such that the expert may wish to define rules based on different types of results.

6.2 Standardisation

It was observed that the knowledge acquisition workload is increased when inconsistent nomenclature is allowed, such as it so often is in many medical systems. To prevent this increased workload for the expert, it would be prudent to derive and enforce a strict scheme for the data input. A possible complication is that users may find it difficult to locate options which are not named as expected. To handle this it would be possible to implement another interpretive layer of hierarchy, essentially allowing the user to use their own preferred nomenclature, and then defining within the system that their chosen nomenclature is synonymous to whichever standardised equivalent is selected by the system designers.

References

1. Peterson, G., *Continuing evidence of inappropriate medication usage in the elderly*, in *Australian Pharmacist*. 2004. p. 2.
2. Bates, D., et al., *Incidence of adverse drug events and potential adverse drug events. Implications for prevention. ADE Prevention Study Group.* JAMA, 1995(274): p. 29-34.
3. Peterson, G., *The future is now: the importance of medication review*, in *Australian Pharmacist*. 2002. p. 268-75.
4. Rigby, D., *The challenge of change - establishing an HMR service in the pharmacy*, in *Australian Pharmacist*. 2004. p. 214-217.
5. Kinrade, W., *Review of Domiciliary Medication Management Review Software*. 2003, Pharmacy Guild of Australia. p. 77.
6. Aamodt, A. and E. Plaza, *Case-Based Reasoning: Foundational Issues, Methodological Variations, and System Approaches*, in *AICom - Artificial Intelligence Communications*. 1994. p. 39-59.
7. Compton, P., et al. *Knowledge Acquisition without Analysis*. in *Knowledge Acquisition for Knowledge-Based Systems*. 1993. Springer Verlag.
8. Tenni, P., et al. to I. Bindoff. 2005.
9. Bonner, C., *MediFlags*. 2005.
10. Compton, P. and R. Jansen. *A philosophical basis for knowledge acquisition*. in *European Knowledge Acquisition for Knowledge-Based Systems*. 1989. Paris.
11. Rivest, R., *Learning Decision Lists*, in *Machine Learning*. 1987. p. 229-246.
12. Kang, B., P. Compton, and P. Preston, *Multiple Classification Ripple Down Rules*. 1994.
13. Kolodner, J., R. Simpson, and K. Sycara-Cyranski. *A Process Model of Cased-based Reasoning in Problem Solving*. in *International Joint Conference on Artificial Intelligence*. 1985. Los Angeles: Morgan Kaufmann.
14. Kolodner, J.L., *Special Issue on Case-Based Reasoning - Introduction*. Machine Learning, 1993. **10**(3): p. 195-199.
15. Kang, B. and P. Compton, *A Maintenance Approach to Case Based Reasoning*. 1994.
16. Kang, B., P. Compton, and P. Preston. *Multiple Classification Ripple Down Rules: Evaluation and Possibilities*. in *AIII-Sponsored Banff Knowledge Acquisition for Knowledge-Based Systems*. 1995. Banff.
17. Bindoff, I., *An Intelligent Decision Support System for Medication Review*, in *Computing*. 2005, University of Tasmania: Hobart. p. 65.
18. Preston, P., G. Edwards, and P. Compton. *A 2000 Rule Expert System Without a Knowledge Engineer*. in *AIII-Sponsored Banff Knowledge Acquisition for Knowledge-Based Systems*. 1994. Banff.
19. Compton, P. and R. Jansen. *Cognitive aspects of knowledge acquisition*. in *AAAI Spring Consortium*. 1992. Stanford.
20. Tenni, P. to I. Bindoff. 2005.

A Hybrid Browsing Mechanism Using Conceptual Scales

Mihye Kim[1] and Paul Compton[2]

[1] Department of Computer Science Education,
Catholic University of Daegu, 712-702, South Korea
mihyekim@cu.ac.kr
[2] School of Computer Science and Engineering,
University of New South Wales, Sydney 2052, Australia
compton@cse.unsw.edu.au

Abstract. A Web-based document management and retrieval system has been developed aimed at small communities in specialized domains and based on free annotation of documents by users. In the proposed approach, the main search mechanism is based on browsing a concept lattice of Formal Concept Analysis (FCA) formulated with a set of keywords with which users annotated the documents. In this paper, we extend our search mechanism by combining the lattice-based browsing structure with conceptual scales of FCA for ontological domain attributes. Our experience with a prototype suggests that conceptual scaling helps users not only to get more specific search results, but also to search relevant documents by the interrelationship between the keywords of documents and ontological attributes.

Keywords: Conceptual scaling, Browsing mechanism, Formal concept analysis.

1 Introduction

Formal Concept Analysis (FCA) was developed by Rudolf Wille in 1982 [17]. FCA is a theory of data analysis which identifies conceptual structures among data based on the philosophical understanding of a 'concept' as a unit of thought comprising its extension and intension as a way of modeling a domain. The extension of a concept is formed by all objects to which the concept applies and the intension consists of all attributes existing in those objects. This results in a lattice structure, where each node is specified by a set of objects and the attributes they share. The mathematical formulae of FCA can be considered as a machine learning algorithm which can facilitate automatic document clustering. A key difference between FCA techniques and general clustering algorithms in Information Retrieval is that the mathematical formulae of FCA produce a concept lattice which provides all possible generalization and specialization relationships between object sets and attribute sets. This means that a concept lattice can represent conceptual hierarchies, which are inherent in the data of a particular domain.

A. Hoffmann et al. (Eds.): PKAW 2006, LNAI 4303, pp. 132–143, 2006.
© Springer-Verlag Berlin Heidelberg 2006

FCA has been successfully applied to a wide range of applications. A variety of methods for data analysis and knowledge discovery in databases have also been proposed based on FCA. A number of researchers have proposed an FCA lattice structure for document retrieval [2], [10], [14]. Several researchers have also studied the lattice-based information retrieval with graphically represented lattices for specific domains such as flight information, e-mail management and real estate advertisements [4], [5], [7]. Recently, FCA has been also applied to ontology engineering for structuring and building of ontologies [3], [15].

We also proposed a theoretical framework for a Web-based document management and retrieval system aimed at small communities in specialized domains based on FCA [13]. This approach allowed users themselves to freely annotate their documents. Any relevant documents can be managed by annotating with any terms the users or authors prefer. A number of annotation support tools were proposed not only to allow users to find appropriate annotations for their documents but also to be able to evolve a terminological domain ontology. This resulted in the automatic generation of a lattice-based browsing system which holds hierarchical inheritance relationships among the evolved terms (concepts) in the lattice structure. Document retrieval is based on navigating this lattice structure.

Experiments were conducted in the domain of annotating researchers' home pages according to their research interests in the School of Computer Science and Engineering, University of New South Wales (UNSW). The goal was a system to assist prospective students and potential collaborators in finding research (i.e., staff and student home pages) relevant to their interests. Results indicated that the annotation tools provided a good level of assistance so that documents were easily organized and a lattice-based browsing structure that evolves in an *ad hoc* fashion provided good efficiency in retrieval performance. It was also clear from the results that there is an advantage in lattice-based browsing over hierarchical browsing. The findings suggested that the concept lattice of FCA, supported by annotation techniques was a useful way of supporting the flexible open management of documents required by individuals, small communities and in specialized domains.

In our approach, the main search mechanism is based on browsing a concept lattice of FCA formulated with a set of keywords with which users annotated documents. This concept lattice is reformulated dynamically and incrementally by the addition of a new document with a set of keywords or by refining the existing keywords of the documents. The concept structure can fit into a predetermined terminological ontology used for browsing in information retrieval.

In this paper, we extend our search mechanism by combining the lattice-based browsing structure with conceptual scales of FCA for ontological information. The purpose of this is to allow a user to get more specific search results and to reduce the complexity of the visualization of the browsing structure. The more fundamental purpose of this is to support a hybrid browsing mechanism by combining a structure with keywords and a structure with ontological attributes. This is to allow a user to search relevant documents by the interrelationship between the keywords of documents and ontological domain attributes. The properties such as *author*, *title* and *publication year* can be ontological attributes in a domain relevant to papers. The ideal would be to support both approaches simultaneously because the organization of background knowledge, not only with the vocabularies in taxonomies but also with ontological structures in the form of properties, would be useful for navigating document. For example, a user may want to find papers which are related to "knowledge acquisition" at first. Then, the user wants to see recently published papers only among the search result (i.e., "publication year" ≥ 2005).

This paper is organized as follows. Section 2 gives a brief description of FCA. Section 3 presents a formal framework of conceptual scaling to combine the lattice-based browsing structure with conceptual scales of FCA for ontological attributes. Section 4 describes a system implemented on the Web to demonstrate the value of conceptual scaling. We then conclude with a discussion of possible future directions of the research presented in this paper.

2 Formal Concept Analysis

2.1 Basic Notions of Formal Concept Analysis

FCA starts with a formal context which is a binary relation between a set of objects and a set of attributes. It was defined for the document retrieval system that we proposed in the paper [13] as follows: A *formal context* is a triple $C = (D, K, I)$ where D is a set of documents (objects), K is a set of keywords (attributes) and I is a binary relation which indicates whether k is a keyword of a document d. If k is a keyword of d, it is written dIk or $(d, k) \in I$.

For the domain of research interests used for experiments in the previous work [13] and used in this paper, a document corresponds to a home page and a set of keywords is a set of research topics. That is, D is the set of home pages and K is the set of research topics for a context (D, K, I). However, the word *documents* and *keywords* are also used interchangeably to denote *home pages* (or simply *pages*) and *research topics* (or simply *topics*), respectively in this paper.

From the formal context, formal concepts and a concept lattice are formulated. A formal concept consists of a pair with its extent and intent. The extension of a concept is formed by all objects to which the concept applies and the intension consists of all attributes existing in those objects. These generate a conceptual hierarchy for the domain by finding all possible formal concepts which reflect a certain relationship between attributes and objects. The resulting subconcept-superconcept relationships between formal concepts are expressed in a concept lattice which can be seen as a semantic net providing "hierarchical conceptual clustering of the objects... and a representation of all implications between the attributes" [18, pp.493]. The implicit and explicit representation of the data allows a meaningful and comprehensible interpretation of the information. This lattice is used as the browsing structure. Fig. 1 shows an example of a lattice and a data structure for organizing documents in the lattice. More detailed formulae and explanations of FCA can be found in [9], [13].

2.2 Conceptual Scaling

Conceptual scaling has been introduced in order to deal with many-valued attributes [8], [9]. Usually more than one attribute exists in an application domain and each attribute may have a range of values so that there is a need to handle many values in a context.

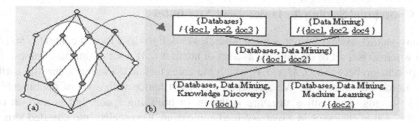

Fig. 1. An example of a browsing structure. (a) Lattice structure. (b) Indexing of the lattice.

In addition, often there is a need to analyze (or interpret) concepts in regard to interrelationships between attributes in a domain. This is the main motivation for conceptual scaling.

For instance, the domain of a "used car market" consists of a number of attributes such as *price*, *year built*, *maker*, *color*, *transmission* and others, and each attribute with a set of values. Such attributes can be considered all together in a context named with a many-valued context. Then, when one is interested in analyzing "used cars" regarding an interrelationship between certain attributes in the many-valued context, they can combine the attributes of interest into a concept lattice. This means that each attribute, or a combination of more than one attribute of the many-valued context, can be transformed into a one-valued context. The derived one-valued context is called a conceptual scale. Then, if one is interested in analyzing the interrelationship between attributes, s/he can choose and combine the conceptual scales which contain the required attributes. This process is called conceptual scaling. A case for the use of this can be seen with TOSCANA[11] and [4]. Conceptual scaling is also used with one-valued contexts in order to reduce the complexity of the visualization [5], [16]. In this case, scales are applied for grouped vertical slices of a large context.

3   Formal Framework of Conceptual Scaling

There are two ways in which we use conceptual scales. Firstly, ontological attributes can be used where readily available (e.g., person, academic position, research group and so on). These correspond to the more structured ontological properties used systems such as Ontoshare[6] and CREAM[12]. The key point of our approach is flexible evolving ontological information but there is no problem with using more fixed information if available. We have included such information for interest and completeness in conceptual scaling. Secondly, a user or a system manager can also group a set of keywords used for the annotation of documents. The groupings are then used for conceptual scaling.

The main difference between our approach and conceptual scaling in TOSCANA is that in our approach all the existing ontological attributes are scaled up together in the nested structure. On the other hand, in the TOSCANA system only one attribute (i.e., a scale) can be combined into the outer structure of an attribute at a time.

3.1   Conceptual Scaling for Ontological Attributes

A many-valued context for ontological attributes is defined as a formal context $C = (D, M, W, I)$ where D is a set of documents, M a set of attributes, W a set of attribute values. I is a ternary relation between D, M and W which indicates that an document d has the attribute value w for the attribute m. We formulate a concept lattice with a set of documents and their keywords. This lattice structure is the main browsing space, but is also an outer structure. Other attributes in a many-valued context are then scaled into a nested structure of the outer structure at retrieval time.

Table 1 is an example of a many-valued context in the domain of research interests. Researchers can be the objects of the context as they are the instances of the home pages. The attributes in the many-valued context can be represented in a partially ordered hierarchy as shown in Fig. 2. The attribute *"position"* in Table 1 is located as a subset of the attribute *"person"* in the hierarchy.

Table 1. An example of the many-valued context for the domain of research interests

	Research group	Sub-group of AI	Person	Position
Researcher1	Artificial intelligence	Knowledge Acquisition	Academic staff	Professor
Researcher2	Computer systems	-	Research staff	Research associate
Researcher3	Networks	-	Academic staff	Associate professor
Researcher4	Databases	-	Academic staff	Senior lecturer
Researcher5	Software engineering	-	Research student	Ph.D. student

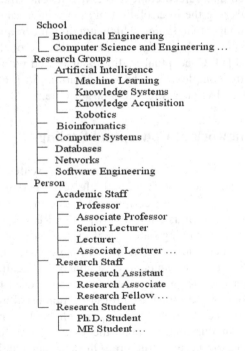

Fig. 2. Partially ordered multi-valued attributes for the domain of research interests

To explain this in a more formal way, the following definition is provided. For example, the has-value relation $\mathfrak{R}$ on the attributes *"person"* and *"position"* is: $\mathfrak{R}$ = {(academic staff, professor), (academic staff, associate professor), ..., (research staff, research assistant), ..., (research student, Ph.D. student), (research student, ME student)} from Fig. 2. This hierarchy of the many-valued context with the relation $\mathfrak{R}$ is scaled into a nested structure using pop-up and pull-down menus.

Definition 1. Let S_p be a super-attribute and S_c be a sub-attribute. There is a binary relation $\mathfrak{R}$ called the "has-value" relation on S_p and S_c such that $(p, c) \in \mathfrak{R}$ where $p \in S_p$ and $c \in S_c$ if and only if c is a sub-attribute value of p.

Fig. 3 shows examples of inner browsing structures corresponding to concepts of the outer lattice. A nested structure is constructed dynamically from the extent (home pages) of a corresponding concept of the outer lattice incorporating the ontological hierarchy. When a user assigns a set of topics for their page, the page is also automatically annotated with the values of the attributes in the many-valued context. A default home page for individual researchers is provided on the School Web site and as well as every researcher has a login account at the School. We make use of this login account when a user annotates their home page. This provides the default home page address of the user.

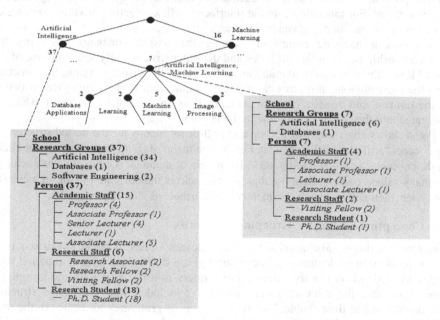

Fig. 3. Examples of nested structures corresponding to concepts. This shows the outer structure of the concepts "artificial intelligence" and "artificial intelligence, machine learning" constructed from a set of home pages and their topics. Numbers in the lattice and in brackets indicate the number of pages corresponding to the concept of the lattice and the attribute value, respectively. The nested structure is presented in a hierarchy deploying all embedded inner structures. The structure is implemented using pop-up and pull-down menus as shown in Fig. 4.

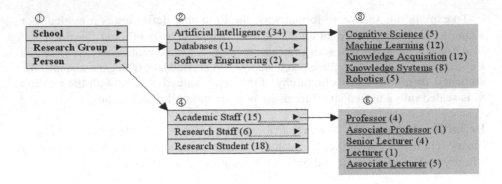

Fig. 4. An example of pop-up and pull-down menus for the nested structures of a concept

The page is an HTML file in a standard format including the basic information about the researcher such as their first name, last name, e-mail address, position and others. The system parses the HTML file and extracts the values for the pre-defined attributes. From the attributes and their extracted values, we formulate a nested structure for a concept of the lattice at retrieval time. Note that the attributes which do not exist in the default home page can be used for conceptual scaling. The user will need to provide the values of those attributes when they assign a set of keyword for their document. For this case, a simple interface to click selection of values or a series of text boxes to be filled is given to the user.

A user can navigate recursively among the nested attributes observing the interrelationship between the attributes and the outer structure. By selecting one of the nested items, the user can moderate the cardinality of the display. Again, the structure with the most obvious attributes can be partly equivalent to the ontological structure of the domain and consequently is considered as an ontological browser which is integrated into the lattice structure with the keywords set.

Fig. 4 shows an example of pop-up and pull-down menus for the nested structure of the concept "artificial intelligence" in Fig. 3. The menu of ① appears when a user clicks on the concept "artificial intelligence". Each item of menu ① is equivalent to a scale in the many-valued context. Suppose that the user selects the attribute *Person* in menu ①, the system then will display a sub-menu of the attribute as shown in menu ④.

3.2   Conceptual Scaling for Grouping Keywords

Conceptual scaling is also applied to group relevant values in the keyword sets used for the annotation of documents. The groupings are determined as required, and their scales are derived on the fly when a user's query is associated with the groupings. This means that the relevant group name(s) is included into the nested structure dynamically at run time. Table 2 shows examples of groupings for scales in the one-valued context for the attribute '*keyword*'. To deal with grouping for scales, the following definition is provided:

Definition 2. Let a formal context $C = (D, K, I)$ be given. A set $G \subseteq K$ is a set of grouping names (generic terms) of C if and only if for each keyword $k \in K$, either $k \in G$ or there exists some generic term $\kappa \in G$ such that k is a sub-term of κ. We define $S = K \setminus G$ and a relation $gen \subseteq G \times S$ such that $(g, s) \in gen$ if and only if s is a sub-term of g.

Table 2. Examples of grouping for scales in the one-valued context for the attribute '*keyword*'

Grouping (generic) names	The members of the grouping names
RDR	FRDR, MCRDR, NRDR, SCRDR
Sisyphus	Sisyphus-I, Sisyphus-II, Sisyphus-III, Sisyphus-IV, Sisyphus-V
Knowledge acquisition	Knowledge acquisition methodologies, Knowledge acquisition tools, Incremental knowledge acquisition, Automatic knowledge acquisition, Web based knowledge acquisition, ...
Computer programming	Concurrent programming, Functional programming, Logic programming, Object oriented programming, ...
Programming languages	Concurrent languages, Knowledge representaion languages, Logic languages, Object oriented languages, ...
Databases	Deductive databases, Distributed databases, Mobile databases, Multimedia databases, Object oriented databases, Relational databases, Spatial databases, Semistructural databases
...	...

Then, when a user's query is $qry \in G$, a *sub-formal context* $C' = (D', K', I')$ of (D, K, I) is formulated where $K' = \{k \in K \mid k = qry$ or $(qry, k) \in gen\}$, $D' = \{d \in D \mid \exists k \in K'$ and $dIk\}$ and $I' = \{ (d, k) \in D' \times K' \mid (d, k) \in I\} \cup \{ (d, qry) \mid d \in D'$ and $qry \in K' \cap G\}$. For instance, suppose that there are groupings as shown in Table 2 and a user's query "*databases*". The query *databases* $\in G$ so that a sub-context C' is constructed to include a scale of the grouping name *databases* and build a lattice of C'. The user can then navigate this lattice of C'.

Fig. 5 shows an example of a scale with the grouping name "*databases*". The grouping name is embedded into an item of the nested structure along with other scales from the many-valued context in the previous section. There are 12 documents with the concept "Databases" in the lattice, and the node (Databases, 12) embeds the scales as shown in menu ①. The scale "Databases" was derived from the groupings in the one-valued context, while other scales (items) were derived from the many-valued context (i.e., ontological attributes). A user can read that there is one document related to "deductive databases", and two documents with "multimedia databases" etc. By selecting an item of sub-menu ②, the user can moderate the retrieved documents which are only associated with the selected sub-term.

A knowledge engineer/user can set up or change the groupings using a supported tool (i.e., ontology editor) whenever it is required. When a grouping name with a set of sub-terms is added, the system gets the set of documents that are associated with at least one of the sub-terms of the grouping name. Then, the context C is refined to have a binary relation between the grouping term and the documents related to the sub-terms of the grouping term. Next, the lattice of C is reformulated when any change in C is made. If a grouping name is changed, it is replaced with the changed one in the context C and its lattice.

In the case of removal of a grouping in the hierarchy, no change is made in the context C. With this mechanism, the outer lattice can always embed a node which can assemble all documents associated with the sub-terms of a grouping. That is, the.

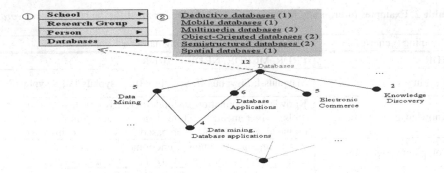

Fig. 5. A conceptual scale for the grouping name "*databases*"

groupings play the role of intermediate nodes in the lattice to scale the relevant values. Groupings can be formed with more than one level of hierarchy. This means that a sub-term of a grouping can be a grouping of other sub-terms.

4 Implementation

To examine the value of conceptual scaling, a prototype has been implemented with a test domain for research topics in the School of Computer Science and Engineering, UNSW. There are around 150 research staff and students in the School who generally have homepages indicating their research projects. The aim here was to allow staff and students to freely annotate their pages so that they would be found appropriately within the evolving lattice of research topics.

Fig. 6 shows an example of conceptual scaling for ontological attributes. It shows examples of inner browsing structures corresponding to the concept "*Artificial Intelligence*" of the outer lattice. We scale up ontological attributes into an inner nested structure. The nested structure is constructed dynamically and associated with the current concept of the outer structure. In other words, the nested attribute values are extracted from the result pages. A nested pop-up menu appears when the user clicks on the "nested" icon in the front of the current node. If the user clicks on one of the attributes items, the results will be changed according to the selection. The user can navigate recursively among the nested attributes.

For instance, we suppose that the user selects the attributes items *Position* → *Academic Staff* → *Professor*. The result then will be changed accordingly. The user can see that there are four researchers whose research topic is "*Artificial Intelligence*" and whose position is *Professor*. Numbers in brackets indicate the number of documents (i.e., homepages) corresponding to the attribute value.

As well, a knowledge engineer can arrange related terms by accessing a tool which allows him or her to set up hierarchical grouping related terms under a common name as described in Section 3.2. Then, when a user's query is related to the grouping(s), the grouping name is included into the inner structure on the fly. Fig. 7 shows an example of conceptual scaling for the grouping "*Databases*". Other items (i.e., *School, Research Groups, Position*) are derived from the ontological attributes. There are 12 documents with the concept "*Databases*" in the lattice.

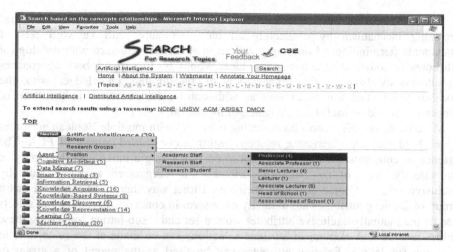

Fig. 6. An example of conceptual scaling for ontological attributes

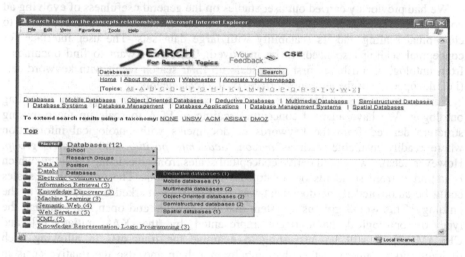

Fig. 7. An example of conceptual scaling for the grouping "*Databases*"

The user can read that there is one document related to "*Mobile databases*", two documents with "*Multimedia databases*" and so on. By selecting one of the grouping members, the user can moderate the retrieved documents which are only associated with the selected sub-term.

5 Discussion and Conclusion

Having completed a prototype implementation of the presented approach, it seems clear that conceptual scaling facilitates users to get more specific results and to search relevant documents by the interrelationship between the keywords of documents and the domain attributes.

Another purpose of conceptual scaling in our approach was to support a hybrid browsing mechanism by connecting an outer structure with keyword sets of documents (terminological ontology) and an inner nested structure with ontological attributes (ontological structure). The ideal would be to support both approaches simultaneously because the organization of background knowledge, with the vocabularies in taxonomies as well as with ontological structures in the form of properties, would be useful for navigating information.

More fundamentally, conceptual scaling is to deal with multiple Boolean attributes which hold multiple inheritance relations within a one-valued context of FCA. The essence of conceptual scaling is to impose on this a single inheritance hierarchy or equivalently some of the Boolean attributes are reorganized as being mutually exclusive values of some unnamed attributes. Either way there is recognition that a group of Boolean attributes are mutually exclusive. In conceptual scaling, one selects one of the mutually exclusive attributes from a set and a sub-lattice containing these values is shown. A number of attribute selections can be made at the same time to give the sub-lattice. Existing attributes can be used as the parent of a group of mutually exclusive attributes or new names for the grouping can be created.

We had previously carried out user studies on the general usefulness of evolving ad hoc lattices [13]. A next step is to evaluate the usefulness of our approach to conceptual scaling and its scalability with large data sets. The user interface for conceptual scaling also needs to be improved. Users may want to find documents from ontological attributes first, then scale up their search result with keyword sets (i.e., the opposite of the current interface) or interchangeably.

Further work would be related to the extension of our approach regarding ontologies. We have adapted conceptual scaling of FCA to scale up the browsing structure derived from the keywords of documents with ontological information where readily available such as *person, academic position* and *research group*. However, ideally we would derive conceptual scales from an existing ontology, which is imported from standards or constructed for the system. The use of these scales could be automated if the document was appropriately marked up according to the ontology. This would give us a system that was flexible and open, but also had the type of ontological commitment represented by the KA^2 initiative[1] and CREAM[12]. It will be interesting to examine the trade-offs in allowing such requirements to emerge rather than anticipating them and also the relative costs in marking up documents rather than providing information to a server.

It will be also essential to use one of ontology representation languages such as RDF, OIL and OWL as standards instead of the proprietary text formats used currently both for the concept lattice and the ontological attributes.

References

1. Benjamins, V. R., Fensel, D., Decker, S., Perez, A. G.: $(KA)^2$: Building Ontologies for the Internet: a Mid-term Report, International journal of human computer studies (1999) 51(3): 687-712.
2. Carpineto, C., Romano, G.: Information retrieval through hybrid navigation of lattice representations, International Journal of Human-Computer Studies (1996) 45:553-578.

3. Cimiano, P., Hotho, A., Stumme, G., Tane, J.: Conceptual Knowledge Processing with Formal Concept Analysis and Ontologies, Proceedings of the Second International Conference on Formal Concept Analysis (ICFCA 04), Springer-Verlag (2004) 189-207.
4. Cole, R., Eklund, P.: Browsing Semi-structured Web texts using Formal Concept Analysis, Conceptual Structures: Broadening the Base, Proceedings of the 9th International Conference on Conceptual Structures (ICCS 2001), Springer-Verlag (2001) 290-303.
5. Cole, R., Stumme, G.: CEM - A Conceptual Email Manager, Conceptual Structures: Logical, Linguistic, and Computational Issues, Proceedings of the 8th International Conference on Conceptual Structures (ICCS 2000), Springer-Verlag (2000) 438-452.
6. Davies, J., Duke, A., Sure, Y.: OntoShare – A Knowledge Environment for Virtual Communities of Practice, Proceedings of the Second International Conference on Knowledge Capture (K-CAP 2003), ACM, New York (2003) 20-27.
7. Eklund, P., Groh, B., Stumme, G., Wille, R.: A Contextual-Logic Extension of TOSCANA, Conceptual Structures: Logical, Linguistic, and Computational Issues, Proceedings of the 8th International Conference on Conceptual Structures (ICCS 2000), Darmstadt, Springer-Verlag (2000) 453-467.
8. Ganter, B., Wille, R.: Conceptual Scaling, In: F. Roberts (eds.): Application of Combinatorics and Graph Theory to the Biological and Social Sciences, Springer-Verlag (1989) 139-167.
9. Ganter, B., Wille, R.: Formal Concept Analysis: Mathematical Foundations, Springer-Verlag, Heidelberg (1999).
10. Godin, R., Missaoui, R., April, A.: Experimental comparison of navigation in a Galois lattice with conventional information retrieval methods, International Journal of Man-Machine Studies (1993) 38 :747-767.
11. Groh, G., Strahringer, S., Wille, R.: TOSCANA-Systems Based on Thesauri, Conceptual Structures: Theory, Tools and Applications, Proceedings of the 6th International Conference on Conceptual Structures (ICCS'98), Springer-Verlag (1998) 127-138.
12. Handschuh, S., Staab, S.: CREAM – CREAting Methadata for the Semantic Web, Computer Networks (2003) 242:579-598.
13. Kim, M., Compton, P.: Evolutionary Document Management and Retrieval for Specialised Domains on the Web, International journal of human computer studies (2004) 60(2): 201-241.
14. Priss, U.: Lattice-based Information Retrieval, Knowledge Organisation (2000) 27(3):132-142.
15. Quan, T.T., Hui, S.C., Fong, A.C.M., Cao, T.H.: Automatic Generation of Ontology for Scholarly Semantic Web, The Semantic Web – ISWC 2004: Proceedings of the Third International Semantic Web Conference, Hiroshima, Springer-Verlag (2004) 726-740.
16. Stumme, G.: Hierarchies of Conceptual Scales, 12th Banff Knowledge Acquisition, Modelling and Management (KAW'99), Banff, Canada, SRDG Publication, University of Calgary (1999) 5.5.1-18.
17. Wille, R.: Restructuring lattice theory: an approach based on hierarchies of concepts, In: Ivan Rival (eds.), Ordered sets, Reidel, Dordrecht-Boston (1982) 445-470.
18. Wille, R.: Concept lattices and conceptual knowledge systems, Computers and Mathematics with Applications (1992) 23:493-515.

Knowledge Representation for Video Assisted by Domain-Specific Ontology

Dan Song[1], Miyoung Cho[1], Chang Choi[1], Juhyun Shin[1],
Jongan Park[2], and Pankoo Kim[3,*]

[1] Dept. of Computer Science
Chosun University, 375 Seosuk-dong Dong-Ku Gwangju 501-759 Korea
songdan@stmail.chosun.ac.kr, irune@chosun.ac.kr,
eduranceaura@gmail.com, jhshinkr@chosun.ac.kr
[2] Dept. of Information and Communication Engineering,
Chosun University, Korea
japark@chosun.ac.kr
[3] Dept. of CSE, Chosun University, Korea
pkkim@chosun.ac.kr

Abstract. Video analysis typically has been pursued in two different directions. Either previous approaches have focused on low-level descriptors, such as dominant color, or they have focused on the video content, such as person or object. In this paper, we present a video analysis environment not only to bridge these two directions but also can extract and manage semantic metadata from multimedia content autonomously for addressing the interaction between browsing and search capabilities. Concretely speaking, we implemented a tool that links MPEG-7 visual descriptors to high-level, domain-specific concepts. Our approach is ontology-driven, in the sense that we provide ontology based domain-specific extensions of the standards for describing the knowledge of video content. In this work, we consider one shot (episode) in the billiard game of video as the specific domain and we will be through the practical works to explain the process of representation of video knowledge. In the experiment part, we prove our approach effectiveness by comparing with the video content retrieval based on only key-word.

1 Introduction

Although new multimedia standards, such as MPEG-4 and MPEG-7 [1], provide the needed functionalities in order to manipulate and transmit objects and metadata, their extraction, and that most importantly at a semantic level, is out of the scope of these standards and is left to the content developer. Extraction of low-level features and object recognition are important phases in developing multimedia database management systems.

There has been a research focus to develop techniques to annotate the content of images on the Web using Web ontology languages such as RDF and OWL. Past efforts have largely focused on mapping low-level image features to ontological

---

* Corresponding author.

A. Hoffmann et al. (Eds.): PKAW 2006, LNAI 4303, pp. 144–155, 2006.

concepts [2] and have involved the development of tools that are closely tied to do-main specific ontologies for annotation purposes [3,4]. Additionally, the lack of pre-cise models and formats for object and system representation and the high complexity of multimedia processing algorithms make the development of fully automatic semantic multimedia analysis and management systems a challenging task. This is due to the diffi-culty that often mentioned as the semantic gap. The use of knowledge domain is probably the only way by which higher level semantics can be incorporated into techniques that capture the semantic concepts. So, in this paper, a comprehensive method for video con-tent analysis based on the specific knowledge domain was proposed using on the tools of Protégé which is the classical ontology editor and PhotoStuff that is the most promising annotation software that allows users to makeup of an image/video key-frame with respect to concepts in an ontology.

We organize the remainder of the paper as follows: Section 2 is about the overview for video analysis. Section 3 introduces the infrastructure of domain knowledge. As the major part, section 4 shows us how to present video content through one specific domain ontology. It contains two sub-sections: ontology building and mapping from the low-level features to high-level semantics for video knowledge representation. And, Analysis results for video content retrieval are showed in Section 5. After these com-prehensive explanations, we will conclude in section 6.

2   Overview for Video Analysis

Video is a structured medium in which actions and events in time and space convey stories, so, a video program (raw video data) must be viewed as a document, not a non-structured sequence of frames.

From Figure 1, we can see the second layer: video conceptual feature which was represented by video shots that are the basic units used for accessing video and a sequence of frames recorded contiguously and re-presenting a continuous action in

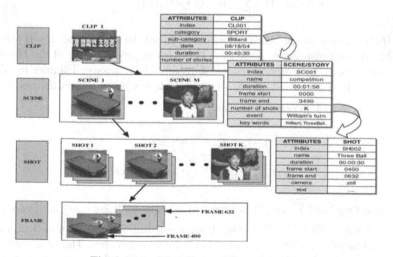

Fig. 1. Video Modeling and Representation

time or space. And we consider a shot that contains a series of actions that can be used to express one meaningful event in the video as one Knowledge Domain.

Since there are three frame types (I, P, and B) in a MPEG bit stream, we first propose a technique to detect the scene cuts occurring on I frames, and the shot boundaries obtained on the I frames are then refined by detecting the scene cuts occurring on P and B frames. For I frames, block-based DCT is used directly as

$$F(u,v) = \frac{C_u C_v}{4} \sum_{x=0}^{7} \sum_{y=0}^{7} I(x,y) \times \cos\frac{(2x+1)u\pi}{16} \cos\frac{(2y+1)v\pi}{16} \qquad (1)$$

Where

$$C_u, C_v = \begin{cases} \frac{1}{\sqrt{2}} & \text{for } u,v = 0 \\ 1 & \text{otherwise} \end{cases} \qquad (2)$$

One finds that the dc image [consisting only of the dc coefficient ($u=v=0$) for each block] is a spatially reduced version of I frame. For a MPEG video bit stream, a sequence of dc images can be constructed by decoding only the dc coefficients of I frames, since dc images retain most of the essential global information of image components.

Yeo and Liu have proposed a novel technique for detecting shot cuts on the basis of dc images of a MPEG bit stream, [5] in which the shot cut detection threshold is determined by analyzing the difference between the highest and second highest histogram difference in the sliding window. In this article, an automatic dc-based technique is proposed which adapts the threshold for shot cut detection to the activities of various videos. The color histogram differences (HD) among successive I frames of a MPEG bit stream can be calculated on the basis of their dc images as

$$HD(j, j-1) = \sum_{k=0}^{M} [H_{j-1}(k) - H_j(k)]^2 \qquad (3)$$

where $H_j(k)$ denotes the dc-based color histogram of the jth I frame, $H_{j-1}(k)$ indicates the dc-based color histogram of the ($j-1$)th I frame, and k is one of the M potential color components. The temporal relationships among successive I frames in a MPEG bit stream are then classified into two opposite classes according to their color histogram differences and an optimal threshold $\overline{T_c}$,

$$HD(j, j-1), > \overline{T_c}, \quad shot\_cut,$$
$$HD(j, j-1), \le \overline{T_c}, \quad non\_shot\_cut \qquad (4)$$

The optimal threshold $\overline{T_c}$ can be determined automatically by using the fast searching technique given in Ref. [5]. The video frames ~including the I, P, and B frames. Between two successive scenes cuts are taken as one video shot. The following figures have shown us the shot we have detected using the algorithm mentioned above. The shot has contains a series of I frames.

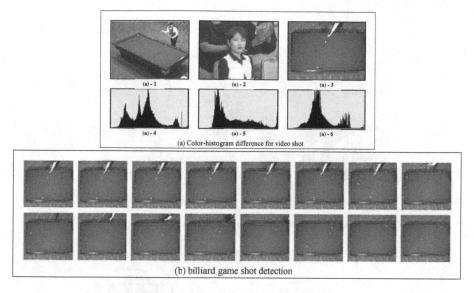

(a) Color-histogram difference for video shot

(b) billiard game shot detection

Fig. 2. Knowledge domain obtainment

3 Domain Knowledge Infrastructure

Video Domain knowledge is usually presented by the visualization in still images and videos in terms of low-level features and media structure descriptors. Structure and semantics are carefully modeled to be largely consistent with existing multimedia description standards like MPEG-7. MPEG-7 is a means of attaching metadata to multimedia content [6], it offers a comprehensive set of audiovisual description tools including metadata elements and their structures and relationships defined by the standard in the form of Descriptors and Description Schemes. The DDL(Description Definition Language) also allows the extension for specific applications of particular DSs [7]. The description tools are instantiated as descriptions in textual format (XML) based on the DDL (based on XML Schema).

Figure 3 illustrates possible conceptual aspects and abstractions of a specific instance (image: "billiard_ShotI01.jpg") of a video shot content. The Structure DSs and Semantic DSs can be related by a set of links allowing the shot content to be described on the basis of both content structure and semantic structures. The links relate different semantic concepts to the instances within the shot content described by the segments.

Furthermore, most of the MPEG-7 content description and content management DSs are linked together and in practice, also often included within each other in the MPEG-7 descriptions. In our case, our structure DS is the event –"Three Cushion", and it related to three kinds of DSs: "Semantic time", "Semantic Location", "Object" and with their corresponding semantic concept definitions. Based on the MPEG-7's Visual Part and Description Scheme, we have created the following billiard game ontology like the framework shows in Figure 4.

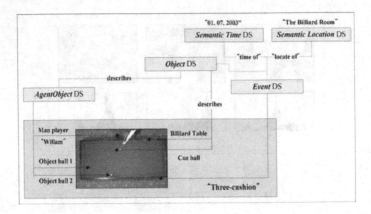

Fig. 3. Conceptual abstractions of a video shot content

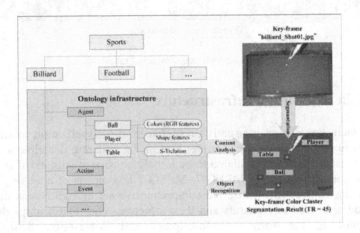

Fig. 4. Ontology based video content representation

In order to achieve our aforementioned aims for bridging the chasm existing between the high and low levels, we propose a comprehensive ontology infrastructure which is based on the MPEG-7 scheme that was analyzed above, and details will be described as follows. The summarized knowledge infrastructure can be divided into two major parts.

One is the domain ontology, in the multimedia annotation framework, is meant to model the content layer of multimedia content with respect to specific real-world domains, such as sports events like billiard game which was considered the example in this paper. We want to extract semantic information from one image but without a gap between the high-level concept and low-level features, the domain ontology should be explored. As the figure shows us, the middle part-"Billiard Ontology" in the knowledge domain of sports plays the important role of "mapping". Ontology is structured as the middle in Figure 4 shows. It contains some significant classes like Event,

Action, Agent and so on and their corresponding instances, for example, in the class of Agent, it has the instances like ball, player, and table, etc, also following their property values we called "low level feature".

The other part is to represent how the visual characteristics are associated with a concept [8, 9]. One has to employ several different visual properties depending on the concept at hand. For instance, in the billiard domain as was described in the scenario in the aforementioned section, the billiard ball might be described using its shape (e.g. round), color (e.g. white or red), or in case of video sequences, motion.

4 Video Content Representation Through Domain-Specific Ontology

Based on knowledge infrastructure summarized in the Section 3, this part will be concentrated on introducing the billiard game domain ontology building in Protégé for video content representation.

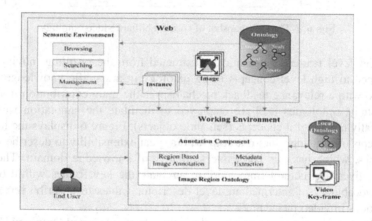

Fig. 5. Video content annotation architecture

First and foremost, video content annotation architecture (depicted in Figure 5) is built in which the Photo-Stuff [10] was considered as the annotating environment. The annotation architecture is primarily composed of two capabilities namely ontology-based video or image annotation and image metadata management on the Semantic Web. We specifically do the experiment towards one specific key-frame in billiard video which was obtained in section 2 through the segmentation algorithm to realize the mapping from low-level feature to semantic level. According to the Architecture shows in Figure 5, we can build our own local ontology for being necessary. By referencing to the MPEG-7 description scheme and instances semantic relations, a billiard game domain ontology was structured in Protégé like the Figure 6 depicted. It shows the Protégé editing environment [11].

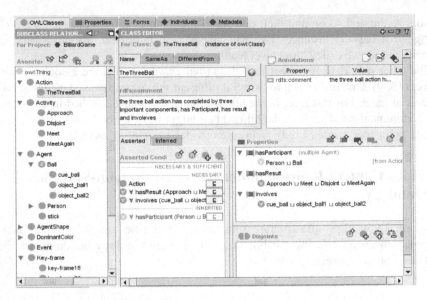

Fig. 6. A Protégé snapshot of specific billiard video ontology

The low-level features automatically extracted from the resulting moving objects are mapped to high-level concepts using ontology in a specific knowledge domain, combined with a relevance feedback mechanism is the main contribution in this part. In this study, ontologies [12] are employed to facilitate the annotation work using semantically meaningful concepts (semantic objects), Figure 6 displays the hierarchical concepts of the video shot ontology with a great extensibility to describe common video clips, which has the distinctive similarity of knowledge domain. The simple ontology gives a structural framework to annotate the key frames within one shot, using a vocabulary of intermediate-level descriptor values to describe semantic objects' actions in video metadata.

The Protégé environment composes these three parts: Asserted Ontology Hierarchy at which we have defined a billiard game domain ontology files in OWL[13]; Class Editor; Asserted Conditions and Properties Definitions areas. OWL classes are interpreted as sets that contain individuals, such as Action, Event, Agent, etc. They are described using formal (mathematical) descriptions that state precisely the requirements for membership of the class. For example, the class "BilliardGame-Agent" would contain all the individuals (ball, player, table, stick and audience, etc.) that are billiard game in our domain of interest. This ontology allows assertions to be made stating that an image contains a region that depicts certain concepts.

Through this domain ontology, we map from the low-level features to the high-level semantic concepts for the video knowledge representation. So in the previous works, a local billiard game ontology has been pre-specified in OWL, defining a small set of concepts for video key-frames, regions, depictions, etc.   And in order to realize the billiard domain ontology's function, we explore the Photostuff software[10] as our assisted tool for video annotation. Photostuff is a platform independent (written in Java), image annotation tool which uses an ontology to provide the expressiveness required to assert the contents of an image, as well information about the image (date created, etc.). The annotation works are proceeded as follows:

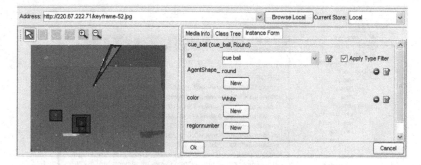

Fig. 7. Specific key-frame(#52) region annotation

Firstly, we load the owl file from our local server which was already defined, then, we import the key-frame(#52) of billiard game video using the local server directory too. When these two elements are well prepared, we begin to annotate the objects that are displayed in the key-frame by specifying the regions of the objects. For example, if we want to annotate the object of "cue ball" and we just choose the rectangle drawing bar to highlight it and the object's corresponding properties will appear on the right side like the figure 7 shows. So, the "instance form" in the figure, we can choose the properties for this object which are already defined in the former ontology, and the same methods to other objects in this key-frame. Therefore, the annotation can be finished semi-automatically assisted by the billiard game ontology.

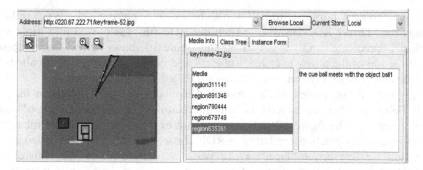

Fig. 8. Specific key-frame(#52) content representation

We specify five significant regions: player, cue_ball, object_ball1, object_ball2, and meet (cue_ball, object_ball1). By associating the cue_ball with the object_ball1's properties, we annotated the fifth region (region635361). We called "the cue_ball meets with the object_ball1". Figure 8 demonstrated the results of key-frame annotation using the action of "meet". The key frame content was annotated with the ontology concepts. After that, we can link the key-frame to the billiard video shot that was obtained from the segmentation in aforementioned section for video event annotation. To simply view the RDF/XML syntax of the annotations, select Windows View RDF(like the figure 9 shows). The RDF output from the image markup performed for key-frame annotation on the Semantic Web.

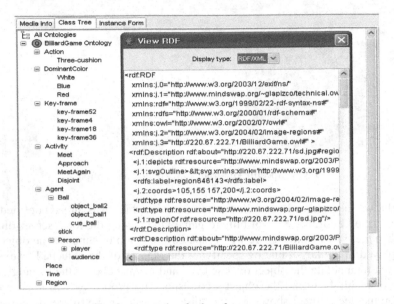

Fig. 9. Annotation for key-frame component

5   Analytical Results for Video Content Retrieval

For video content retrieval, the RDF validation is a prerequisite. We validate it through the "Driver RDF Browser". Then the RDF file was parsed into triple style: "subject", "predicate" and "object". Like the figure 10 shows below, the component lays at the top of the arrow is called "subject" and the bottom of arrow is called "object', the component between them is what we called -" predicate".

We got total 400 triples of data model, 126 nodes and 101 literals in the RDF graph. Following that, we convert them to the relational model based on our "billiard game" ontology. The figure 10 shows a snapshot of a hierarchical model for expression of inheritance-relationship between RDF classes that describes the video content annotation for one action-"meet" which was one participant in the billiard game event-"the three-ball". If we want to find the action - "meet", we should focus on this relational model but not to consider other information such "disjoint" or "audience applause" information that there is of no use for the users' retrieval.

We do the experiment by comparing the ontology-based with key-word retrieval. The following figure 11 shows the process for video retrieval base on billiard game ontology. For example, suppose the user want to retrieve as this: *I want to find "the three cushion" participated by William?*   After parsing, we extract the semantic

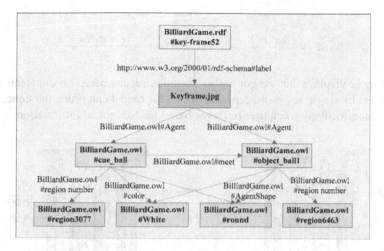

Fig. 10. RDF graph for billiard game key-frame annotation content

information(Player: *William,* Action: *the three cushion*; Parameter mapping: *rdf: Agent, rdf: Event*). Next step is to compare the semantic information parameters with the RDF relation model which contains RDF metadata.

After the retrieval, we evaluate our experiment results, based on this relational model. We adopted an effective method of retrieval which is usually measured by the following two quantities, recall and precision

$$recall = \frac{Number\ of\ retrieval\ relevant\ object}{Number\ of\ relevant\ objects} \quad Precision = \frac{Number\ of\ retrieval\ relevant\ objects}{Number\ of\ retrieval\ objects} \quad (1)$$

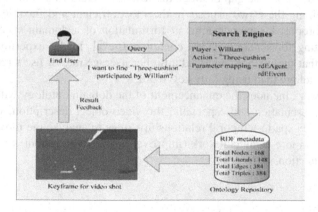

Fig. 11. Ontology- based query processing

A relevant object is an object of use to the user in response to his or her query. Let us assume that *Rel* represents the set of relevant objects and *Ret* represents the set of retrieved objects. The above measure can also be redefined in the following manner.

$$Recall = \frac{|\ Rel\ \cap\ Ret\ |}{|\ Rel\ |} \qquad Precison = \frac{|\ Rel\ \cap\ Ret\ |}{|\ Ret\ |} \qquad (2)$$

The Fig 12 displays that we got higher accuracy results based on our relational mo
del than just by single key-word retrieval does. Our model can prune the unnecessary
information effectively when user retrieves based on conceptual information.

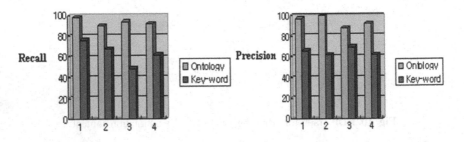

Fig. 12. Recall and precision of ontology-based and key-word based research

6 Conclusions

In this paper, we proposed a novel method for video content analysis and description
on the fundamental of the knowledge domain ontology. In this work we have presented a generic, domain independent framework for annotating and managing digital
image content using the Semantic Web technologies. The adaptive billiard ontology
not only helps us overcome gap between the low-level features and high level semantics, but combining these two aspects in the most efficient and flexible (expressive)
manner. The proposed approach aims at formulation of a domain specific analysis
model facilitating for semantic video content retrieval. In the experimental part, it
demonstrated that the method achieved a higher average rate of use's retrieval based
on semantic ontology than using only key-word.

Our future work includes the enhancement of the domain ontology with more complex model representation and especially, the video object description, we try to use
the more complex spatio-temporal relationships rules to analyze the moving features.
We will also do more technical work (improve our retrieval system) to intensify our
retrieval part function.

References

1. S.-F. Chang, T. Sikora, and A. Puri. Overview of the MPEG-7 standard. IEEE Trans. on
   Circuits and Systems for Video Technology, 11(6):688–695, June 2001.
2. Dupplaw, D., Dasmahapatra, S.,Hu, B., Lewis, P., and Shadbolt, N. Multimedia Distributed Knowledge Management in MIAKT. ISWC Workshop on Knowledge Markup and
   Semantic Annotation.Hiroshima, Japan, November 2004.

3. Hollink, L., Schreiber, G., Wielemaker J., and Wielinga. B. Semantic Annotation of Image Collections. In Proceedings of Knowledge Capture - Knowledge Markup and Semantic Annotation Workshop 2003.
4. Schreiber, G., Dubbeldam, B., Wielemaker, J., and Wielinga, B. Ontology-Based Photo Annotation. IEEE Intelligent Systems, 16(3) 2001.
5. B. L. Yeo and B. Liu, "Rapid scene change detection on compressed video," IEEE Trans. Circuits Syst. Video Technol. 5, 533–544, 1995.
6. M. G. Strintzis S. Bloehdorn S. Handschuh S. Staab N. Simou V. Tzouvaras K. Petridis, I. Kompatsiaris and Y. Avrithis, "Knowledge representation for semantic multimedia content analysis and reasoning," in European Workshop on the Integration of Knowledge, Semantics and Digital Media Technology, November 2004.
7. Multimedia content description interface-part 8: extraction and use of Mpeg-7 descriptors. ISO/IEC 15938-8, 2002.
8. Kompatsiaris, V. Mezaris, and M. G. Strintzis. Multimedia content indexing and retrieval using an object ontology. Multimedia Content and Semantic Web - Methods, Standards and Tools Editor G.Stamou, Wiley, New York, NY, 2004.
9. A.G. Hauptmann. Towards a large scale concept ontology for broadcast video. In Proc. of the 3rd int conf on Image and Video Retrieval (CIVR'04), 2004.
10. ChristianHalaschek-Wiener, NikolaosSimou and VassilisTzouvaras, "Image Annotation on the Semantic Web," W3C Candidate Recommendation, W3C Working Draft 22 March, 2006.
11. Matthew Horridge Holger Knublauch and Alan Rector, "A Practical Guide To Building OWL Ontologies Using The Prot´eg´e-OWL Plugin and CO-ODE Tools Edition 1.0," W3C Candidate Recommendation, 10 August 27, 2004.
12. A. T. Schreiber, B. Dubbeldam, J. Wielemaker, and B. Wielinga, "Ontology-based photo annotation," IEEE Intell. Syst., vol. 16, pp. 66–74,May-June 2001.
13. T.Adanck, N.O'Connor, and N.Murphy. Region-based Segmentation of Image Using Syntactic Visual Features. In Pro. Workshop on Image Analysis for Multimedia Interactive Services, WIAMIS 2005, Montreux, Swizerland, April 13-15 2005.

An Ontological Infrastructure for the Semantic Integration of Clinical Archetypes

Jesualdo Tomás Fernández-Breis[1], Marcos Menárguez-Tortosa[1], David Moner[2], Rafael Valencia-García[1], José Alberto Maldonado[2], Pedro José Vivancos-Vicente[1], Teddy Gonzalo Miranda-Mena[3], and Rodrigo Martínez-Béjar[1]

[1] Departamento de Informática y Sistemas, Facultad de Informática
Universidad de Murcia, CP 30071 Murcia, Spain
{jfernand, valencia, pedroviv, marcos, rodrigo}@um.es
[2] Bioengineering, Electronics and Tele-Medicine Group, Technical University of Valencia
Building 8G, Camino Vera s/n, CP 46022 Valencia, Spain
damoca@gmail.com, jamaldo@upvnet.upv.es
[3] Intelligent Methodologies (IMET)
Paseo Fotógrafo Verdú, n 11, bajo 2 CP 30002 Murcia, Spain
tmiranda@imet.es

Abstract. One of the basic needs for any healthcare professional is to be able to access clinical information of patients in an understandable and normalized way. The lifelong clinical information of any person supported by electronic means configures his Electronic Health Record (EHR). There are currently different standards for representing EHRs. Each standard defines its own information models, so that, in order to promote the interoperability among standard-compliant information systems, the different information models must be semantically integrated. In this work, we present an ontological approach to promote interoperability among CEN- and OpenEHR- compliant information systems by facilitating the construction of interoperable clinical archetypes.

1 Introduction

One of the basic needs for any healthcare professional is to be able to access clinical information of patients in an understandable and normalized way. The lifelong clinical information of any person supported by electronic means configures his Electronic Health Record (EHR). This information is usually distributed among several independent and heterogeneous systems that may be syntactically or semantically incompatible. There are currently different standards for representing electronic healthcare records (EHR). Each standard defines its own information models and manages the information in a particular way. This implies that clinical information systems of different clinical organizations might differ in the way electronic healthcare records are managed. Hence, exchanging healthcare information among health professionals or clinical information systems is nowadays a critical process for the healthcare sector. Due to the special sensitivity of medical data and its ethical and legal constraints, this exchange must be done in a meaningful way,

A. Hoffmann et al. (Eds.): PKAW 2006, LNAI 4303, pp. 156–167, 2006.

avoiding all possibility of misunderstanding or misinterpretation. Two main problems arise when pursuing that objective. On the one hand, many hospitals do not have a unified information system. Health data is distributed across several heterogeneous and autonomous systems whose interconnection and integration is difficult to achieve. This may be resolved by setting up a new and integrated information system for all the organization but it would represent a great economic cost, a traumatic upgrade of existing applications and a difficult adaptation of current users to the new system.

On the other hand, a clinical information system may lack a comprehensive semantic definition of the information which it contains, up to the point of making impossible a semantic interoperability between different systems. Due to the complexity and constant evolution of health domain this has not an easy solution. As stated in [6] not only is medicine domain big, it is open-ended because new information, finer grained details or new relationships are always being discovered or becoming relevant. As a consequence, no fixed enumerated list of medical concepts can ever be complete. This implies that a traditional information model will never be completely adapted to the clinical requirements and its continuous evolution.

The main techniques that have traditionally been applied to obtain integration and interoperability at application level are adaptors and exchange formats, whose success has not been very significant to date. Therefore, alternative approaches are currently making use of semantic technologies to facilitate integration and interoperability [4; 9; 10]. An advantage of using semantic approaches is the fact that they do not require to replace current integration technologies, databases and applications, but add a new layer that takes advantage of the already existing infrastructure [5]. Amongst the different available semantic technologies, ontologies are considered a basic technology to promote semantic interoperability between independent and heterogeneous systems [7]. An ontology, which represents a common, shareable and reusable view of a particular application domain [11], gives meaning to information structures that are exchanged by information systems. This paper presents a semantic approach to facilitate the interoperability of EHR information models. This work has been focused on two EHR standards: CEN ENV13606 and OpenEHR. Both information models have been analyzed and semantically represented by means of ontologies. Then, the ontological models have been integrated into a common ontological infrastructure, which will be the core for developing model-independent systems.

Finally, the structure of this paper is the following. In section 2, models for representing electronic healthcare records are discussed. Section 3 describes the role ontologies can play in integration and interoperability issues. Then, the ontological infrastructure developed for this work is described in section 4. Section 5 describes the process for constructing the ontologies of the different information models. Section 6 contains an example of archetype modelling using the ontological infrastructure. Finally, some conclusions will be put forward.

2 Electronic Healthcare Records

A healthcare record is the set of non-redundant, ordered, and complete information concerning the relation between an individual and any healthcare centre. Healthcare records can be queried in different situations and due to different reasons, and they can play different roles in the healthcare process. Their main use is to support clinical care. In the last years, different working groups have been actively working in the definition of architectures and information models for electronic healthcare records. Each model implies a working environment in which the meaning of data varies. This requirement is fulfilled by semantic technologies, which make the description of the nature and logical context of the information to exchange possible, allowing each system to remain independent. An Electronic Healthcare Record (EHR) is a healthcare record digitally stored in one or more information systems.

The OpenEHR consortium has developed the dual model architecture approach [1] for electronic healthcare records. This architecture is based on the metamodelling of healthcare records, and it is based on the separation of concepts in two levels: (1) reference model (RM), and (2) archetypes, which are formal models of clinical concepts. The information system is based on the RM and the valid healthcare records extracts are instances of this reference model. This methodology was tested in the Good Electronic Healthcare Record project (GEHR) in Australia and by the European Synex project. It is also used in the new version of HL7 and in the CEN norm for the communication of healthcare records.

The reference model represents the global features of the annotations of healthcare records, how they are aggregated and the context information required to meet the ethical, legal, etc requirements. This model defines the set of classes that form the generic building blocks of the electronic healthcare record and it contains the non-volatile features of the electronic healthcare record. However, the reference model needs the complement of domain knowledge: archetypes. An archetype models the common features of types of entities and, therefore, defines the valid domain structures. Archetypes restrict the business objects defined in a reference model, bridging the generality of business concepts defined in the reference model and the variability of the clinical practice. They provide a standard tool to represent this issue. Archetype instances are expressed in an archetype definition language (ADL) and they are therefore related to formal archetype model, which is formally related to the reference model. Although both ADL and archetype models are stable, the individual archetypes can be modified in order to be adapted to clinical practice.

The work developed in projects such as the previously mentioned GEHR and OpenEHR suggest that the formalisms for defining archetypes must be based on the following main technical principles: (1) each archetype is a different and complete domain concept; (2) archetypes are expressed as restrictions on the reference model; (3) the granularity of an archetype corresponds to the granularity of a business concept of the reference model; (4) each business concept can be considered a descriptor of a domain ontological level; and (5) archetypes have partonomic and taxonomic components. Having introduced how electronic healthcare records can be represented at conceptual level, let us describe the two current EHR standards that are based on a dual model approach: CEN/TC251 EN13606 and OpenEHR on which we are focusing in this research work both at information level (reference model) and knowledge model (archetypes).

2.1 CEN

The CEN/TC251, Technical Committee 251 of the Normalization European Committee [12] is in charge of developing standards in the field of medical informatics. The activity of one of its working groups is devoted to the standardization of the architecture and information models for electronic healthcare records. The overall goal of this standard is to define a rigorous and durable information architecture for representing the EHR, in order to support the interoperability of systems and components that need to interact with EHR services: as discrete systems or as middleware components, to access, transfer, add, or modify health record entries, via electronic messages or distributed objects, preserving the original clinical meaning intended by the author, reflecting the confidentiality of that data as intended by the author and patient.

This standard will have five parts: (1) generic information model for communicating the electronic healthcare record of any one patient (reference model); (2) generic information model and language for representing and communicating the definition of individual instances of archetypes (archetype exchange specification:); (3) a range of archetypes reflecting a diversity of clinical requirements and settings, as a "starter set" for adopters and to illustrate how other clinical domains might similarly be represented (reference archetypes and term lists); (4) the information model concepts that need to be reflected within individual EHR instances to enable suitable interaction with the security components (security features); and (5) a set of models built on the above parts and can form the basis of message-based or service-based communication (exchange models).

This model makes use of the dual approach for communicating the electronic healthcare record. Here, an archetype is defined as a computable expression of a clinical domain concept based on a reference model. It is defined through a set of structured restrictions. Archetypes share the same formalism, but they can be of different types. Definitional archetypes are part of a standardized, shared ontology. Non definitional archetypes are locally used and defined by particular institutions to fulfil particular clinical needs. However, clinical organizations should agree on common definitions in order to exchange clinical information efficiently. Therefore, obtaining the mappings between these archetypes might be of interest.

2.2 OpenEHR

The OpenEHR Foundation [13] is a non-profit organization. Amongst its objectives, the following ones can be pointed out: promote and publish the formal specification of requirements for representing and communicating electronic health record information, based on implementation experience, and evolving over time as health care and medical knowledge develop; promote and publish EHR information architectures, models and data dictionaries, tested in implementations, which meet these requirements; and manage the sequential validation of the EHR architectures through comprehensive implementation and clinical evaluation.

The openEHR architecture [2] specifications consist of the following components: (1) Reference Model, which provides identification, access to knowledge resources, data types and structures, versioning semantics, and support for archetyping; (2) Service Model, which defines the basic services in the health information environment, centred around the EHR; and (3) Archetype Model, which describe the semantics of archetypes and templates, and their use within openEHR.

All of the architecture specifications published by openEHR are defined as a set of abstract models. Among the global requirements of EHRs and EHR systems supported by openEHR, the following can be pointed out: life-long EHR; priority to the patient / clinician interaction; technology and data format independent; facilitation of EHRs sharing via interoperability at data and knowledge levels; integration with any/multiple terminologies; support for clinical data structures: lists, tables, time-series, including point and interval events; compatibility with CEN 13606, Corbamed, and messaging systems.

3 Ontologies for Integration and Interoperability

The most important factors that make the integration and interoperability between systems difficult are the semantic and structural heterogeneity, as well as different meaning information has in different systems. Hence, our interest is focused on how semantic technologies, in particular ontologies, may support and promote interoperability among electronic healthcare records systems.

An ontology can be seen as a semantic model containing concepts, their properties, interconceptual relations, and axioms related to the previous elements. Furthermore, ontology has a standard reference model to integrate information known as knowledge sharing. In practical settings, ontologies have become widely used due to the advantages they have (see for instance [3]). On the one hand, ontologies are reusable, that is, a same ontology can be reused in different applications, either individually or in combination with other ontologies. On the other hand, ontologies are shareable, that is, their knowledge allows for being shared by a particular community. In the context of integration, they facilitate the human understanding of the information. Ontologies allow for differentiating among resources, and this is especially useful when there are resources with redundant data. Thus, they help to fully understand the meaning and context of the information. This is important for our objective of achieving semantic interoperability among electronic healthcare record systems built on top of different information models. For our purpose, the information model semantics is formalized by means of ontologies and represented by using the Ontology Web Language (OWL) [18].

Ontologies have already been used for integration and interoperability purposes in medical domains. In [8], ontologies were used to promote integration and interoperability between information systems for three medical communities by combining data with HL7 [14] and terminologies such as UMLS [15], MEDCIN [17] and SNOMED [16]. So, terminologies are integrated by using ontologies. Our approach is different because EHR standards have a different nature and the components defined in clinical archetypes can be linked to different terminologies. Therefore, our work can benefit from terminological integration approaches such as [8] in order to simplify the management of different terminologies at EHR level. Another example is the joint effort made by the ONTOLOG [19] forum, the Medical Informatics department of Stanford University and the Semantic Interoperability Community of Practice (SICOP) [20] to integrate and make the Federal Health Architecture and the National Health Information Network interoperable. They defined an three level ontological architecture. The medium level ontologies were FEA-RMO (Federal Enterprise Architecture- Reference Model Ontology) and HL7

RIM, and different domain ontologies were obtained from the Federal Health Architecture. This effort was carried out in the context of HL7, so different EHR models were not targeted as we do in this work.

4 The Ontological Infrastructure

In this section, the ontological infrastructure to facilitate the interoperability between CEN- and OpenEHR-based information systems is described. In this work, the following versions of the CEN and OpenEHR specifications have been used: CEN (release 09/2004) and OpenEHR (release 09/2005). A set of components will be described, as well as their use and the relations among them. The information models have clear differences related to how information items are organized and taxonomic depth in specific issues. However, they do not present inconsistencies so that the quality mappings between both models are likely to be achieved.

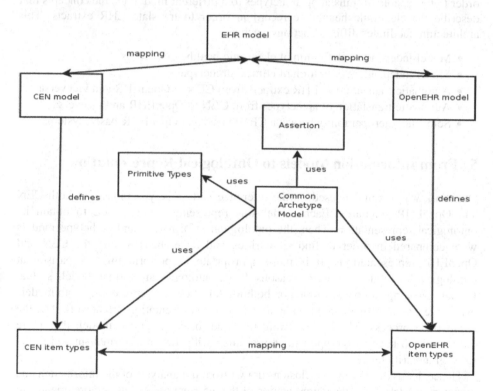

Fig. 1. The ontological infrastructure

Figure 1 shows the set of ontologies that configure our ontological architecture to solve this problem, as well as the relations existing among them. This figure has three main areas. The left and right areas contain specific information of each reference model (i.e, CEN, OpenEHR). Each one defines a healthcare record information model, including a set of types of items. For instance, clusters and elements are types of CEN items. The OpenEHR is richer in terms of types of items. The definition of mappings between types of items is possible, although some semantic processing is needed. These items define the nature of clinical actions that can be represented by clinical archetypes. The central part of the figure represents the common parts to both standards. The global reference model is located on top of the figure. This model will allow us to translate information between both models. The archetype model is located in the central part of the figure. This is common for both standards, and it makes use of the ontologies of assertions and primitive data types, which are semantically equivalent for both standards. Furthermore, the archetype model will be used to build archetypes, and these will be used to generate the archetype instances (one per patient). These instances are contained in the extract of the electronic healthcare record of the patients. The lower part of the figure refers to the types of clinical items defined in both models.

Hence, two major mappings have to be made between: (1) the types of items in order to be capable of translating archetypes to a different model; (2) the concepts that describe the electronic healthcare record in order to translate EHR extracts. This architecture facilitates different actions, such as:

- Model-independent definition of electronic healthcare records
- Model-independent definition of clinical archetypes
- Automatic translation of EHR extracts from CEN to OpenEHR and viceversa
- Automatic translation of archetypes from CEN to OpenEHR and viceversa
- Semantic interoperability between CEN-based and OpenEHR-based systems

5   From Information Models to Ontological Representation

Our work was mainly focused on the reference and archetype models of both CEN and OpenEHR standards. Each model was represented using OWL to obtain its ontological representation. Then, the ontological information and archetype models were compared in order to find similarities and differences among the CEN and OpenEHR representations. It is more appropriate to perform this comparison at ontological level due to different reasons. First, ontologies are formal models so that formal reasoning can be performed on both models. Second, representing both models using the same formalism provide a common representation framework for the comparison process. Third, if we want to come to an integrated model, it is more appropriate to have the components represented with the same formalism and at the same granularity level.

Hence, let us discuss the conclusions drawn from the analysis of the information and archetype models. First, the representation of the information models is more oriented to the transmission via a communication network rather than to representing contents semantically. In fact, the different model diagrams provided in the documentation of both standards has little semantic information; they are similar to UML class diagrams. This can be observed in the archetype model defined in the CEN [12] or OpenEHR documentation [13]. For instance, there are some references to elements belonging to different classes modelled by string attributes. This representation may make it harder to understand the underlying semantics. Therefore, it would be more appropriate to model this reference through a relation between the corresponding classes. In our ontological approach, referential semantics is modelled through semantic relations between the concepts. Moreover, the UML-like representation is not suitable for performing formal reasoning at conceptual level, so that, better use of the information contained in the model might be made. This process was performed in two steps: reference model, and archetype model.

Both archetype models contain the same type of information: translations, audit details descriptions, ontological section, and constraints. The ontological section and the constraint are the most important parts since they contain the definition of the archetype terms, which are instances of some specific type of concepts belonging to the respective (CEN/OpenEHR) reference models. However, the archetype model does not structure appropriately or represent formally this information. So, what we have done is to remodel this archetype model by using as modelling focus the archetype terms. Our first goal was to develop an ontology that would represent an integrated view to the CEN and OpenEHR archetypes models, that is, the commonalities were identified, and the differences were kept to allow building archetypes for both reference models. According to our approach, an archetype has general information as it is specified in the standard models. This general information encompasses the auditory details, the archetype description, assertions, translations to other languages, and a set of available terminologies. Furthermore, an archetype contains the definition of a concept (i.e., heart rate pulse), is a specialization of another archetype, and it contains a set of archetype terms. An archetype translation to a specific language is comprised of the set of

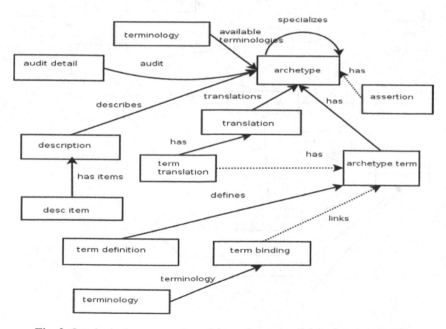

Fig. 2. Ontological representation of the archetype model (archetype concept)

translations of archetype terms to such language. Therefore, each archetype term has a set of translations associated. Each archetype term has also a definition, and a set of term bindings to the available terminologies. Figure 2 shows a part of the ontological representation of the archetype model. This part of the ontology reflects common information to CEN and OpenEHR. Archetype terms can refer to restrictions and conceptual entities. Conceptual terms (called ontology terms) are divided into concepts (e.g., heart rate), complex terms (e.g., list, history), simple terms (e.g., position, device), or values (e.g., sitting, lying). A simple term has a set of values associated. Each complex term is comprised of a set of complex and simple terms. Values are of a particular datatype, which is given by the reference model (CEN / OpenEHR). Both standards use the same basic datatypes, but have different simple and complex terms.

This part of the ontology is shown in Figure 3. Besides the modeling of this integrated archetype model, the reference model of the CEN and OpenEHR standards have been ontologically modeled. For this purpose, the procedure was similar to the one followed for the archetypes model. They were analyzed in order to detect semantic representation flaws and OWL schemes were developed. Then, both ontologies were semantically compared in order to look for mapping between both standards to develop an integrated model for the electronic healthcare records. The main difference is the richness for defining types of clinical data. The CEN model makes use of folders, sections, entries,

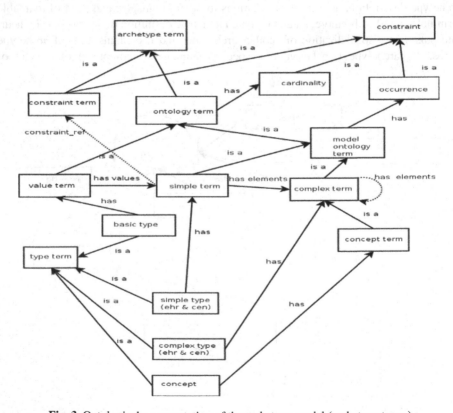

Fig. 3. Ontological representation of the archetype model (archetype terms)

items, clusters and elements, whereas the OpenEHR model uses a wider range of types, including some such as history, item list, item structure, and so on. We are currently completing this mapping, which would allow to transform automatically CEN electronic healthcare records and archetypes into OpenEHR ones and viceversa. Our research group has been working of the modelling of the different standards and definition of the mappings for the last year. The set of OWL ontologies and mappings obtained through this research project are available at http://klt.inf.um.es/~poseacle.

A "problem" has a "structure" and a "protocol". The structure of a problem has a description of the problem, a "date of initial onset", "age at initial onset", "severity", "clinical description", "date clinically recognised", "location", "aetiology", "occurrences or exacerbations", "related problems", "date of resolution", and "age at resolution". These archetype terms must be assigned a type of term. For this purpose, a set of types are available, those belonging to OpenEHR and CEN. Provided that CEN types are a subset of OpenEHR ones, the latter ones can be used as the ones proposed to the archetype builder and then, these can be easily mapped onto CEN types by considering complex OpenEHR types as CEN Cluster and the singles ones as CEN Element. Let us consider the definition of the "protocol". A protocol is comprised of a set of "references", having each a "reference" and a "web link". The following code of the protocol is shown in ADL, where ITEM_TREE, CLUSTER, and ELEMENT refer to the corresponding OpenEHR types., at00xx stands for the terms associated to the entities.

```
protocol matches {
    ITEM_TREE[at0032] matches { -- Tree
        items cardinality matches {0..*; unordered} matches {
            CLUSTER[at0033] occurrences matches {0..1} matches {    -- References
                items cardinality matches {0..*; unordered} matches {
                    ELEMENT[at0034] occurrences matches {0..*} matches {   --
Reference
                        value matches {  TEXT matches {*}      }    }
                    ELEMENT[at0035] occurrences matches {0..*} matches { --
Web link

                        value matches {  URI matches {*}                  }}}}}}}
```

Therefore, our approach would associate an ITEM_TREE to "protocol", a CLUSTER to "references", an ELEMENT to "reference" and to "Web link". These types are mapped onto the same types in OpenEHR. In the case of CEN, the CLUSTER and the ELEMENTs are mapped onto CLUSTER and ELEMENT but the ITEM_TREE is mapped onto a CLUSTER too. In general, any OpenEHR complex type is mapped onto a CLUSTER in CEN. The corresponding occurrences and cardinality constraints are defined similarly in both models.

6 Conclusions

Healthcare professionals need to access the complete healthcare record of their patients in order to perform more efficient healthcare processes. However, this information is usually distributed across heterogeneous sources and systems. Therefore, there is a need for solutions that allow for integrating the information contained in the different sources and systems, and these solutions should ideally be transparent for users. In this paper, this issue has been tackled by applying a semantic approach. Most of the scientific community agrees on the role and importance of the use of semantic technologies. However, a part of the community says that there is currently too much diversity and low standardization in how to work with these technologies. Our semantic approach aims at facilitating the integration and interoperability between CEN and OpenEHR compliant clinical information systems. The key semantic technology to achieve this goal is the ontology, which can be viewed as a conceptual model containing a set of interrelated elements whose existence is accepted by a particular community. Ontologies acquire more importance when they cover particular domains, because once achieved the semantic control of a domain, data integration or linking systems would be easier. In this paper, the effort has been put on generating ontological models of the CEN and OpenEHR reference models, as well as developing an integrated archetype model.

The model proposed in this work would facilitate the interoperability of those information systems that makes use of different EHR models, since the model defines an ontology-based common syntax and semantics. We are currently addressing the mapping between the ontologies corresponding to the CEN and OpenEHR reference models in order to obtain a global EHR model, so that, systems might work with data coming from both standards. The next step will be the development of a model-independent archetype management system, capable of managing both CEN and OpenEHR archetypes by using the ontological infrastructure. This system will provide us qualitative information about this infrastructure.

Acknowledgements

This work has been possible thanks to the Spanish Ministry for Science and Education through the projects TSI2004-06475-C01 and TSI2004-06475-C02; and FUNDESOCO through project FDS-2004-001-01.

References

1. Beale, T.: Archetypes, Constraint-based Domain Models for Future-proof Information Systems (2001) Available: http://www.deepthought.com.au/it/archetypes/ archetypes.pdf
2. Beale, T. and Heard, S. The openEHR Archetype System (2003) Available: http://www.openehr.org/
3. Fernández-Breis, J.T., Martínez-Béjar,R.: A Cooperative Framework for Integrating Ontologies. International Journal of Human-Computer Studies, 56(6) (2002) 662-717
4. Linthicum, D.: Leveraging Ontologies: The Intersection of Data Integration and Business Intelligence Part I . DMR Review Magazine (2004) June.
5. Missikoff, M.: Harmonise: An Ontology-based Approach for Semantic Interoperability. ERCIM News 51 (2002)

6. Rector, A.L:. Clinical Terminology: Why is it so hard?. Methods of Information in Medicine **6** (1999) 245:251.
7. Partridge, C.: The Role of Ontology in Semantic Integration (2002)
8. Paterson, G.I.: Semantic Interoperability for Decision Support Using Case Formalism and Controlled Vocabulary. Health'04, Challenges for Today for Success Tomorrow (2004)
9. Semantic Interoperability Community of Practice: White Paper Series Module 1: Introducing Semantic Technologies and the Vision of the Semantic Web (2005)
10. Stuckenschmidt, H., Wache, H., Visser, U., Schuster, G.: Methodologies for Ontology-based Semantic Translation. ECIMF (2001)
11. Van Heijst, G., Schreiber, A. T., Wielinga, B. J. : Using explicit ontologies in KBS development. International Journal of Human-Computer Studies, **45**, 183-292, (1997).
12. http://www.centc251.org
13. http://www.openehr.org
14. http://www.hl7.org
15. http://www.nlm.nih.gov/research/umls/
16. http://www.snomed.org/
17. http://www.medicomp.com
18. http://www.w3.org/TR/owl-ref/
19. http://ontolog.cim3.net/
20. http://colab.cim3.net/cgi-bin/wiki.pl?SICoP

Improvement of Air Handling Unit Control Performance Using Reinforcement Learning*

Sangjo Youk[1], Moonseong Kim[2], Yangsok Kim[3], and Gilcheol Park[1]

[1] School of Information & Multimedia, Hannam University
133 Ojung-Dong, Daeduk-Gu, Daejeon 306-791, Korea
{youksj, gcpark}@hannam.ac.kr
[2] Dept. Medical Information System, Daewon Science College
599 Shinwol-Dong, Jechon City, Chungbuk 390-702, Korea
kms@mail.daewon.ac.kr
[3] School of Computing, University of Tasmania
Hobart, Tasmania, 7001, Australia
{yangsokk, bhkang}@utas.edu.au

Abstract. Most common applications using neural networks for control problems are the automatic controls using the artificial perceptual function. These control mechanisms are similar to those of the intelligent and pattern recognition control of an adaptive method frequently performed by the animate nature. Many automated buildings are using HVAC(Heating Ventilating and Air Conditioning) by PI that has simple and solid characteristics. However, to keep up good performance, proper tuning and re-tuning are necessary. In this paper, as the one of method to solve the above problems and improve control performance of controller, using reinforcement learning method for the one of neural network learning method(supervised/unsupervised/reinforcement learning), reinforcement learning controller is proposed and the validity will be evaluated under the real operating condition of AHU(Air Handling Unit) in the environment chamber.

1   Introduction

Although modern control theory has been rapidly developing, most industrial controllers of air conditionings and refrigerators, etc use the controller type of PID(Proportional Integral Derivative). In spite of having a simple structure, PID controller is the most used for industrial processor control because it is well-functioned with stability and tenacity of the goal following, invasions from the outside and the process variables. In addition, it experimentally extracts dynamics of the plant by using several tuning methods. It is able to design the controller by searching variable for the optimum control[1,2,3].

However, it cannot estimate control performance in advance if there is uncertainty of process model or a change of operation environment. It is necessary to have an

---

* This paper has been supported by the 2006 Hannam University Research Fund.

A. Hoffmann et al. (Eds.): PKAW 2006, LNAI 4303, pp. 168–176, 2006.

exact estimate of control environment that changes at any time and automatic tuning function, especially, in the case of automatic building control that only operates by typical non-linear structure. In order to choose the parameter of PID controller to get this optimum control function, there have been a lot of studies about tunings of PID controller such as the Ziegler-Nichols[4] tuning in 1942 and the tuning by relayed experiments by Astrom and Hagglund, etc, and a few methods which had improved from the Ziegler-Nicholas tuning are used[5,6,7,8,9].

Therefore, this study has designed the optimum building cooperating controller by using Q-Learning to solve these problems above and improve control function of the controller. Q-Learning was developed by free model reinforcement learning on the based of probable dynamic programming. It has applied to the building cooperating system of artificial climate laboratory inside where it is able to control freely outdoor temperature artificially.

2 Reinforcement Learning

There are actor-critic structure by Sutton's Temporal Difference(TD) and Watkins' Q-learning etc. These methods estimate reinforcement signals and learn when there is no immediate reward at the present. Therefore, reinforcement learning suits the case which exists various models and environments because the controller itself learns the right control signals by using evaluated signals from the environment of controller's behaviours even though a human does not involve the learning and is adaptable like a human even if the environment changes.

The Q-learning method used in this study had developed as one of the methods of free model reinforcement learning which is based on probable dynamic programming. This makes the system with learning ability on the Markovian environment operates the optimum.

In order to use this Q-Learning, if let's say S for a state set and A for an action set of scattered environment, basic Q-Learning algorithm is the same as Fig. 1.

① For all environments s and actions a, $Q(s,a)$ initializes an optional value (generally 0).

② recognizes present environment s.

③ chooses action a according to the rule of environment-action.

④ operates action a on the environment given and then puts s' for the environment and r for immediate reward.

⑤ renews the rule of environment-action from s, a, s' and r.

$$\Delta Q_\pi(s_t, a_t) = \alpha_t[R(s_t, a_t) + \gamma \max_{a' \in A} Q_\pi(s_{t+1}, a') - Q_\pi(s_t, a_t)]$$

α_t : learning rate, γ : decrease factor between 0 and 1

⑥ goes back to step ②.

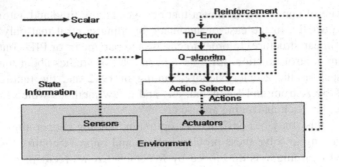

Fig. 1. Reinforcement learning architecture

Fig. 2. shows the summary that combines the structure of reinforcement learning with PI control algorithm. As we studied from Fig. 1. and Fig. 2., reinforcement learning achieves on-line study through two structural elements. The actor by the given environment acts suitable action to the environment and the environment by the given action sends a judgement on that if the changed state and action were right to the actor with reinforcement signal that is a reward for the output of PI controller.

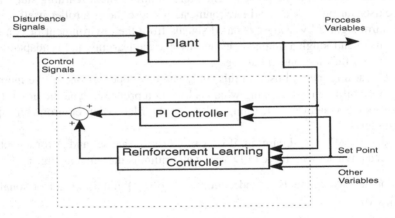

Fig. 2. Combination of RL and PI controller

3 Experimental Devices

3.1 Test House

I built a test house in the artificial climate lab building in order that I could experiment overall such as the load of air-conditioning and heating of the building, the efficiency of air-conditioning and heating, thermal environment, energy saving, heat transfer of the building structure, Wall thermal mass effects, HVAC control, Access floor control, and so on.

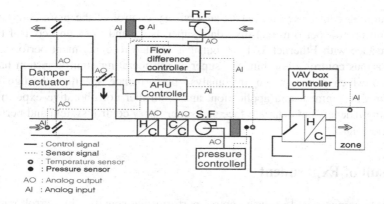

Fig. 3. Configuration of VAV AHU

Fig. 3. shows a composition of cooperating automatic control system which was installed at both the non-hypocaust of the test house and well-equipped lab in the way of variable air volume(VAV). The cooperating system installed was designed to oper ate with a cooperating machine according to the condition of the outside air and in-door. Air-supplier or ventilator are able to do variable voltage variable frequency (VVVF) control, so they can perform economical efficiency and energy saving evaluation. And a variable air volume(VAV) was installed to save energy and control the volume of air which is supplied to each indoor by the load change of inside build-ing. Operations and information collecting about all equipments of the building per-forms in the automatic control system.

3.2 System Embodiment

Supervisory control of cooperating system achieves at the supervisory control of the main computer and local loop control.

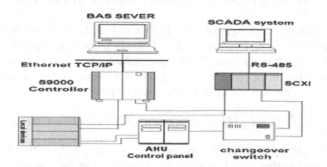

Fig. 4. System realization

Fig. 4. shows the composition of supervisory operating control system for auto-matic operating of cooperation system of the test house. The system is composed of established supervisory operating control system and supervisory operating control system for control algorithm performance experiment separately. Established supervisory operating control system makes supervisory control of the main computer and local loop control perform real-time data supervisory and operating control through data interface with Ethernet TCP/IP, but it is limited to experiment performance of actual various control algorithm. So, supervisory operating control system has been embodied which can compare and analyse performance specifics through control algorithm development and application, and controller corroborative experiment by composing independent data interface and supervisory control system and performing automatic control.

4   Result of Experiment

In order to compare and analyse control performance specifics by corroborative experiment of reinforcement learning controller, it performed performance experiment by using VAV AHU in the test building compared to established PID controller. In the performance experiment, heating coils' the control performance experiment of heating coils for supplied air temperature control has been performed to examine application possibility on the real system before it is applied to whole system. Here are the conditions for the experiment: the temperature of the outside air was -1℃ ~ 0℃, temperature change of mixed air temperature of the test building and supplied air temperature was 22℃ < Tma(temperature of mixed air) < 28℃, 33℃ < Tsa(temperature of supply air) <43℃. With these conditions, the system was operated and it performed the controller performance experiment. Before performance experiment, in order for the system to operate more stable and précised control, it used the optimum control variable of PI controller, that are a comparison element Kp, an integral element Ki, by using the tuning method of Ziegler - Nichols[3] and testing various types of loops. Reinforcement learning controller was added to PI controller, then it used the PI controller that has a comparison element Kp=1.9, an integral element Ki=7.5 to remove heating coils and designed the controller by operating control performance experiment and using output reward control signal of reinforcement learning controller.

It decides 7 of scattered output signals: [-2, -1, -0.2, 0, 0.2, 1, 2] as output reward control signals of reinforcement learning(RL) and sends them with output control signals of PI controller. Each input variable of 3 divided into 8 of space limits and 3-demention input space has been fixed 7^3(343). Each input space stores 7 scattered input signals at Q-value. This is a reinforcement learning equation as below.

$$R(t) = (T_{sa}(t)^* - T_{sa}(t))^2 \quad T_{sa}(t)^* : \text{a set point}, \quad T_{sa}(t) : \text{a observed value}$$

Fig. 5. shows the control performance specific of the heating coils of the case that it sent a random optional output signal according to input state in the condition of RL controller added to PI controller for the online study for the RL controller.

Fig. 6. presents the control performance specific of the heating coils both in the case of PI controller which has the optimum control variable: Kp = 1.9, Ki = 7.5 with little overshoot through the tuning process and in the case of RL controller added to PI

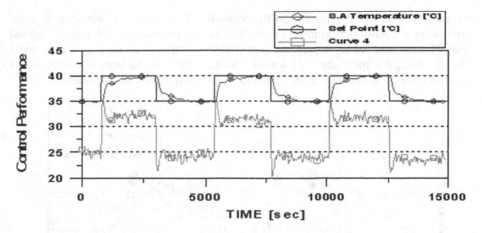

Fig. 5. Control performance when the selected action of RL agent is added to PI controller

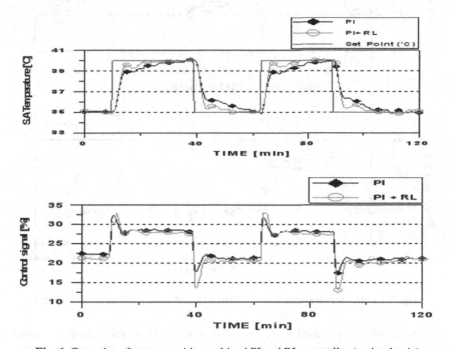

Fig. 6. Control performance with combined PI and RL controller (optimal gain)

controller after completing learning. The optimum controller with complete learning, that is, if RL controller is used, there is a big decrease of normal state error compared to PI controller when a set point for the supply air temperature changes according to the change of outside environment, and an improvement of controller's performance that has quick respond.

Fig. 7. shows the control performance specific of the heating coils both in the case of using PI controller which has control variables Kp= 1, Ki= 4 with increased rising time a bit and the case of applying PI and RL controller. It is clear to be seen there is a improvement of control performance like normal state error decrease and quicker respond than PI controller when supply air temperature changes according to environment change. Also, the more times of learning routine, the less performance lowering of learning errors, it is certain that the optimum control of heating coils is possible.

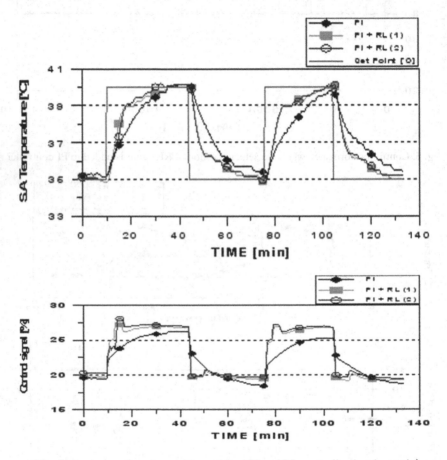

Fig. 7. Control performance with combined PI and RL controller (random gain)

Fig. 8. shows when it used a RL controller with learning error and a PI controller, if you compare the performance, the respond is a bit improved. However, normal state error increased far more than using only PI controller because shaking status quo on the output control signal of heating coils happened by the matter of convergence near the boundary value of the input variable for learning.

To apply the controller which RL controller is applied to actual system through this experiment, first, there would be enough learning, appreciate selection of input variables according to environment change, a boundary range for each variable, and effective selection of output reward signal value of each system to develop the controller with optimum performance.

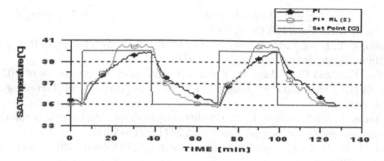

Fig. 8. Performance comparison of controller as learning error

5 Conclusion

As the result of experiment that PI controller and Reinforcement Learning controller applied to actual tuning system, the more increasing times of learning routine, RL controller increases or decreases output of PI controller when the temperature value of supply air changes according to the change of outside environment. It does not decrease normal state error between a set point and a temperature of supply air. However, it is clear that it has a rather quick respond so that output control signal of heating coils improves considerably as it following-controls a set point.

In order to apply the controller which RL controller connects to actual system, there should be enough learning, appropriate selection of input variables, space range of each variable and establishment of control signal value of output reward of each system effectively. Otherwise, there could cause a place where the boundary section condition of variables or no learning achieved, or would not converge because of the happening of shaking status from learning error. Therefore, it could cause deterioration of controller's performance. However, it could be the controller with optimum performance on the purposes of each system if there would be appropriate input variables, reward output signals, and the range of learning times.

The controller with Reinforcement Learning control algorithm has a quick adaptability to environment changes so that it is able to improve the performance of controller.

References

1. Virk, G. S. and Loveday, D. L.: A Comparison of Predictive, PID, and On/Off Techniques for Energy Management and Control, Proceedings of ASHRAE (1992) 3-10.
2. Åström, K. J. and Hägglund, T.: PID controllers: Theory, design and tuning, Research-Triangle Park, NC: Instrument Society of America (1995)
3. Hang, C. C. and Åström, K. J. and Ho, W. K.: Ziegler-Nichols tuning formula., Vol. 138. IEE Proc. D, No.2, 111-118.
4. Ziegler, J. G. and Nichols, N. B.: Optimum settings for automatic controllers, Trans. ASME (1942) 433-444
5. Hang, C. C., LIM, C. C. and Soon, S. H.: A new PID auto-tunner design based on correlation technique, Proc. 2nd Multinational Instrumentation Conf., China(1986)

6. Hang, C. C. and Åström, K. J.: Refinements of the Ziegler Nichols tunning formula for PID auto-tunners, Proc. ISA Conf., USA
7. Åström, K. J., Hang, C. C., Persson, P., Ho, W. K.: Towards Intelligent PID Control, International Federation of Automatic Control(1991)
8. Åström, K. J. and C. C. Hang and P. Persson.: Heuristics for assessment of PID control with Ziegler-Nichols tuning, Automatic Control, Lund Institute of Technology, Lund, Sweden(1988)
9. Åström, K. J., and Hagglund, T.: Automatic tuning of simple regulators with specifications on phase and amplitude margins, Automatica 20 (1984) 645-651.
10. Sutton, R. S.: Learning to predict by the methods of TD(temporal differences). Machine Learn. 3, (1988) 9-44.
11. Anderson, C. W.: Q-learning with hidden-unit restarting, In Advances in Neural information processing system 5, (1993) 81-88.
12. Barto. A. G. and Bradtke, S. J. and Singh, S. P.:, Learning to act using real-time dynamic programming, Artificial Intelligence 72, (1995) 81-138.
13. Watkins, C. J. and DAYAN, P.: Q-learning, Machine Learning 8 (1992) 279-292.

Optimizing Dissimilarity-Based Classifiers Using a Newly Modified Hausdorff Distance*

Sang-Woon Kim

Senior Member IEEE, Dept. of Computer Science and Engineering,
Myongji University, Yongin, 449-728 Korea
kimsw@mju.ac.kr

Abstract. The aim of this paper is to present a dissimilarity measure strategy by which a new philosophy for pattern classification that pertaining to Dissimilarity-Based Classifiers (DBCs) can be efficiently implemented. DBCs, proposed by Duin and his co-authors, is not based on the feature measurements of the individual patterns, but rather on a suitable dissimilarity measure between them. The advantage of DBCs is that since it does not operate on the class-conditional distributions, the accuracy can exceed the Bayes' error bound. The problem with this strategy, however, is that we need to measure the inter-pattern dissimilarities for all the training samples such that there is no zero distance between objects of different classes. Consequently, the classes do not overlap, and therefore, the lower error bound is zero. Thus, to achieve the desired classification accuracy, a suitable method of measuring dissimilarities is required to overcome the limitations based on the object variations. In this paper, to optimize DBCs, we suggest a newly modified Hausdorff distance measure, which determines the distance directly from the input gray-level image without extracting the binary edge image from it. Also, instead of obtaining the Hausdorff distance on the basis of the entire image, we advocate the use of a spatially weighted mask, which divides the entire image region into several subregions according to their importance. For instance, in face recognition, important regions could include eyes and mouth, while the rest is considered unimportant regions. There could also be the background region that contains no facial parts. The present experimental results, which, to the best of the authors' knowledge, are the first reported results, demonstrate that the proposed mechanism could increase the classification accuracy when compared with the "conventional" approaches for a well-known face database.

1   Introduction

One of the most recent and novel developments in the field of statistical Pattern Recognition (PR) [1] is the concept of Dissimilarity-Based Classifiers (DBCs) proposed by Duin and his co-authors (see [2], [3], [4], [5], [7]). Philosophically,

---

* This work was generously supported by the KOSEF, the Korea Science and Engineering Foundation (F01-2006-000-10008-0).

A. Hoffmann et al. (Eds.): PKAW 2006, LNAI 4303, pp. 177–186, 2006.

the motivation for DBCs is the following: If we assume that "Similar" objects can be grouped together to form a class, the "class" is nothing more than a set of these "similar" objects. Based on this idea, Duin and his colleagues argue that the notion of proximity (similarity or dissimilarity) is actually more fundamental than that of a feature or a class. Indeed, it is probably more likely that the brain uses an intuitive DBC-based methodology rather than that of taking measurements, inverting matrices etc. Thus, DBCs are a way of defining classifiers between the classes, which are not based on the feature measurements of the individual patterns, but rather on a suitable *dissimilarity measure* between them. The advantage of this methodology is that since it does not operate on the class-conditional distributions, the accuracy can exceed the Bayes' error bound and actually attempt to attain the zero-error bound[1] - which is, in our opinion, remarkable. Another salient advantage of such a paradigm is that it does not have to confront the problems associated with feature spaces, such as the "curse of dimensionality" [1], and the issue of estimating a large number of parameters.

However, DBCs have several problems to be solved when being applied for particular tasks, such as face recognition [11]. The questions encountered in designing the DBCs are summarized as follows: (1) How to select (or create) prototype subsets from the training samples. (2) How to measure the dissimilarities between object samples. (3) How to design a classifier in the dissimilarity space. The existing strategies that have been investigated to answer the above questions are described in the following.

First of all, using *all* of the input vectors as prototypes is a simple way to select prototype subsets. However, in most cases, this will impose a computational burden on the classifier. Recently, Duin and his colleagues [3], [7] discussed a number of methods including *Random, Random_C, KCentres*, where a training set, T, is pruned to yield a set of representative *prototypes*, Y, where, without loss of generality, $|Y| \leq |T|$. Additionally, by invoking a Prototype Reduction Scheme (PRS), Kim and Oommen [9] also obtained the representative subset, Y, which is utilized by the DBC. Apart from utilizing PRSs, in [9], they have also proposed simultaneously employing the Mahalanobis distance as the dissimilarity-measurement criterion to increase the DBC's classification accuracy.

Secondly, regarding the second question, investigations have focused on measuring the appropriate dissimilarity using various L_p Norms (including the Euclidean and $L_{0.8}$), the Hausdorff and Modified Hausdorff norm, traditional PR-based measures, such as those used in Template matching, and Correlation-based analysis. The final question refers to the learning paradigms, especially those which deal either with non-metric or non-Euclidean. Since dissimilarity representations are interpreted in all vector spaces, the tools available for the feature representations may be used for learning from the dissimilarity representation as well. The literature [5], [6] report the use of many traditional decision classifiers

---

[1] The idea of the zero-error bound is based on the fact that dissimilarities may be defined such that there is no zero distance between objects of different classes. Consequently the classes do not overlap, and so the lower error bound is zero. We are grateful to Bob Duin for providing us with insight into this.

including k-NN rule and the linear/quadratic normal-density-based classifiers in the task of classifying z using $\delta_Y(z)$ in the dissimilarity space.

The major task of this study is to deal with how the dissimilarity measure can be effectively computed. The reason why this task is set as our goal comes from the necessity to measure the inter-pattern dissimilarities for (in the worst case) all the training samples such that there is no zero distance between objects of different classes. Consequently, the classes do not overlap, and therefore, the lower error bound is zero [8]. Traditionally, the Hausdorff distance measure have been successfully used in the areas of object recognition. This technique has the advantages of simplicity in algorithm and efficiency in computation, but it is sensitive to degradation, such as noise and occlusions. Thus, numerous modified versions have been proposed in the literature [18], [19], [20].

Recently, Guo et al.[19] proposed a modified Hausdorff distance measure, namely, spatially Weighted Hausdorff Distance (WHD), for face recognition. In [19], the modified Hausdorff distance was weighted according to a weighted function derived from the spatial information of human face based on the fact that different facial regions have different degrees of significance for face recognition. However, WHD also computes the Hausdorff distance from binary images. To reduce the sensitivity associated with the binary images, Zhao et al.[20] proposed another modified Hausdorff distance measure in which the binary image is transformed into a gray image, and the Hausdorff distance is then computed for the gray image.

In this paper, we propose to utilize a newly modified Hausdorff distance as a method of measuring the resemblance of two gray images to construct the dissimilarity matrix of DBCs. The newly modified Hausdorff distance, the spatially Weighted Gray-level Hausdorff Distance (WGHD), determines the distance directly from the input gray-level image without extracting the binary edge image from it to overcome the problems based on the variability of the facial images. Also, instead of obtaining the Hausdorff distance on the basis of the entire image, we advocate the use of the spatially weighted mask, which divides the image region into several subregions according to their importance. For instance, the entire region could comprise the important region, such as eyes and mouth, and the rest being the unimportant region. There could also exist the background region that contains no facial parts.

The modest contributions are claimed in this paper by the authors:

1. This paper proposes a new strategy through which the Hausdorff distance is incorporated with a weighting function so that distances at the regions of important facial features can be emphasized. This modified Hausdorff distance, namely, the spatially Weighted Gray-level Hausdorff Distance (WGHD), can measure the dissimilarity directly from the input gray-level images without extracting the binary edge images.

2. This paper lists the first reported results that improve the performance of DBCs by utilizing the newly modified Hausdorff distance, namely, WGHD, to construct the dissimilarity matrix directly from the gray-level images. Although the result presented is only for the case of face recognition, the proposed approach

can also be applied to other classification tasks, such as information retrieval or text classification.

This paper is organized as follows: An overview on the dissimilarity-based classification is initially presented in Section 2. Following this, in Section 3, a newly modified Hausdorff distance, namely, WGHD, that measures the dissimilarity between two gray-level images is presented. Experimental results for the real-life benchmark data set are provided in Section 4, and the paper is concluded in Section 5.

2   Dissimilarity-Based Classification

Foundations of DBCs: Let $T = \{x_1, \cdots, x_n\} \in R^p$ be a set of n feature vectors in a p-dimensional space. We assume that T is a labelled data set so that T can be decomposed into, for example, c disjoint subsets $\{T_1, \cdots, T_c\}$ such that $T = \bigcup_{k=1}^{c} T_k, T_i \cap T_j = \phi, \forall i \neq j$. Our goal is to design a DBC in an appropriate dissimilarity space constructed with this *training data* set and to classify an input sample z into an appropriate class. To achieve this, we assume that from T_i, the training data of class ω_i, $T_i = \{x_1, \cdots, x_{n_i}\}, n = \sum_{i=1}^{c} n_i$. Then, we extract a prototype set[2], Y_i, where, $Y_i = \{y_1, \cdots, y_{m_i}\}, m = \sum_{i=1}^{c} m_i$.

Every DBC assumes the use of a dissimilarity measure, d, computed from the samples, where $d(x_i, y_j)$ represents the dissimilarity between two samples, x_i and y_j. The measure, d, is required to be nonnegative, reflexive and symmetric[3], and thus, $d(x_i, y_j) \geq 0$ with $d(x_i, y_j) = 0$ if $x_i = y_j$, and $d(x_i, y_j) = d(y_j, x_i)$. The dissimilarity computed between T and Y leads to a $n \times m$ matrix, $D_{T,Y}[i, j]$, where $x_i \in T$ and $y_j \in Y$. Consequently, an object x_i is represented as a column vector as following : $[d(x_i, y_1), d(x_i, y_2), \cdots, d(x_i, y_m)]^T$, $1 \leq i \leq n$. Here, we define the dissimilarity matrix $D_{T,Y}[\cdot, \cdot]^4$ as a *dissimilarity space* on which the p-dimensional object, x, given in the feature space, is represented as an m-dimensional vector $\delta(x, Y)$, where if $x = x_i$, $\delta(x_i, Y)$ is the ith row of $D_{T,Y}[\cdot, \cdot]$. In this paper, the column vector $\delta(x, Y)$ is simply denoted by $\delta_Y(x)$, where the latter is an m-dimensional vector, while x is p-dimensional.

Dissimilarity Measures Used in DBCs: Fundamental to DBCs is the measure used to quantify the dissimilarity between two vectors[5]. The work in [7] reports extensive experiments conducted using various dissimilarity measures (see Table 2 of [7]). The measures which were tested in [7] essentially fall into three categories : (a) The City Block, $L_{0.8}$, Euclidean, and Max Norm, which are special cases of the L_p metric for $p = 1, 0.8, 2$ and ∞ respectively, (b) The

---

[2] Rather Y_i may be created or selected from T_i, and its computation may also involve the other sets, $T_j, j \neq i$.

[3] Note that $d(\cdot, \cdot)$ need not be a *metric* [7].

[4] Here, the subscripts of D represent the set of elements on which the dissimilarities are evaluated.

[5] The details of the binary, categorical, ordinal, symbolic and quantitative features are omitted here, but can be found in [7].

Hausdorff Norm and its variants, which involve *Max-Min* computations, and (c) Traditional pattern recognition norms, such as the Template matching and Correlation Norms. The details of the other measures, such as the *Median* and *Cosine*, are omitted here in the interest of compactness, but can be found in [5].

Classifying Steps in DBCs: Based on what has been discussed above, in all brevity, we state that the state-of-the-art strategy applicable for optimizing DBCs involves the following steps:

1. Select the representative set, Y, from the training set T by resorting to a selection method, for example, the *RandomC* or the *KCentres* algorithm.

2. Compute the dissimilarity matrix $D_{T,Y}[\cdot, \cdot]$, in which each individual dissimilarity is computed using one of the measures described in Section 2.

3. For a testing sample z, compute a dissimilarity column vector, $\delta_Y(z)$, by using the same measure used in Step 2.

4. Achieve the classification by invoking a classifier built in the dissimilarity space and operating it on the dissimilarity vector $\delta_Y(z)$.

From the above steps, we can observe that the performance of DBCs relies heavily on how well the dissimilarity space, which is determined by the dissimilarity matrix $D_{T,Y}[\cdot, \cdot]$, is constructed. Eventually, to improve the performance, the dissimilarity matrix should be well designed, and how to do so is presented in the next section.

3   Spatially Weighted Gray-Level Hausdorff Distance

The Classical Hausdorff Distance: The Hausdorff distance is a *max-min* distance that measures the extent to which two images are similar or different to one another based on their edge maps. Given two finite sets of points $A = \{a_1, \cdots, a_p\}$ and $B = \{b_1, \cdots, b_q\}$, the classical Hausdorff distance is defined as follows:

$$H(A, B) = max\{h(A, B), h(B, A)\}, \tag{1}$$

where the directed distance $h(A, B)$ is:

$$h(A, B) = \max_{a \in A} d(a, B) = \max_{a \in A} \min_{b \in B} \|a - b\|, \tag{2}$$

where $\| \cdot \|$ is an underlying norm on the points of A and B [6]. Then, the directed distance from B to A, $h(B, A) = \max_{b \in A} d(b, A)$, can be computed in the same way as in Eq. (2).

The Modified Hausdorff Distances: The classical Hausdorff distance is simple, but it is sensitive to degradation, such as noise and occlusions. Thus, a number of modified Hausdorff distances have been proposed. Recently, Guo et al. [19] proposed a modified Hausdorff distance, namely, the spatially Weighted

---

[6] The distances, such as the city block, chessboard, and Euclidean distances, are the most popular in pattern recognition and image processing applications. In this paper, we employed the Euclidean distance.

Hausdorff Distance (WHD), where a weighted function was defined according to the various regions with different importance in classification. This WHD is defined as follows:

$$H(A, B) = max\{h_{WHD}(A, B), h_{WHD}(B, A)\}, \tag{3}$$

where the directed distance $h_{WHD}(A, B)$ is:

$$h_{WHD}(A, B) = \frac{1}{N_a} \sum_{N_a} w(a) \min_{b \in B} \|a - b\|, \tag{4}$$

where N_a is the number of points in set A; $w(x)$ is a weighted function, whose definition is:

$$w(x) = \begin{cases} 1 & x \in R_i \\ w_v(0 < w_v < 1) & x \in R_u \\ 0 & x \in R_b \end{cases} , \tag{5}$$

where R_i represents the important facial regions, such as the eyes and mouth, which should be emphasized; R_u is the unimportant facial regions; and R_b is the background regions. The $h_{WHD}(B, A)$ of Eq. (3) is also computed in the same way.

Additionally, Zhao et al. [20] proposed a new method in which the binary image is transformed into the gray image, and then Hausdorff distance is computed for the gray image with a purpose of reducing the intensity of the random noise. The direct distance from A to B in the method is defined as follows:

$$h(A, B) = \max_{a_t \in A} d(a_t, B), \ (1 \le t \le 8), \tag{6}$$

where t is the number of gray-levels; and

$$d(a_t, B) = \min \left(\min_{b_{t-1} \in B} \|a_t - b_{t-1}\|, \min_{b_t \in B} \|a_t - b_t\|, \min_{b_{t+1} \in B} \|a_t - b_{t+1}\| \right). \tag{7}$$

In this method, the authors of [20] compute not only the distance between the point a_t in the finite point set A and the same value b_t in the finite point set B, but also the distance between the a_t and its two neighbor values b_{t-1} and b_{t+1} in the finite point set B. Then, these three distances are minimized.

A Newly Modified Hausdorff Distance: It is well known that edge detection is an ill-posed problem by itself, and the definition of edge is usually ambiguous. To avoid this problem, in this paper, a newly modified Hausdorff distance, namely, the spatially Weighted Gray-level Hausdorff Distance (WGHD), is proposed. In WGHD, the Zhao's distance mentioned in the previous section is incorporated with the Guo's weighting function so that the spatially weighted Hausdorff distance can be measured directly from input gray images.

The direct distance from A to B in the proposed WGHD is defined as follows:

$$h_{WGHD}(A, B) = \frac{1}{N_{a_t}} \sum_{a_t \in A} w(a_t) \, d(a_t, B), \tag{8}$$

where $\sum_{a_t \in A} w(a_t) = N_{a_t}$; and

$$d(a_t, B) = \min \left(\min_{b_{t-k} \in B} \|a_t - b_{t-k}\|, \min_{b_{t-k+1} \in B} \|a_t - b_{t-k+1}\|, \cdots, \right. \tag{9}$$

$$\left. \min_{b_{t-k} \in B} \|a_t - b_{t-k}\|, \cdots, \min_{b_{t+k} \in B} \|a_t - b_{t+k}\| \right).$$

In Eq. (9), k is the number of gray-levels, which are in the neighborhood of the gray-level t. For instance, $k = 2$ means that one needs to compute the distances between the point a_t and its four neighbor values, b_{t-2}, b_{t-1}, b_{t+1}, and b_{t+2}, as well as the distance between the point a_t and the same value b_t, and then minimize these five distances.

In order to apply the above Hausdorff distance, WGHD, to the construction of the dissimilarity matrix of DBCs, an input image is first translated into a multi-level gray image. This translation can be performed by invoking a thresholding algorithm, such as Wu's multi-level thresholding algorithm [21], where the number of thresholding levels to separate the image into segmented ones is determined by measuring the separability of the homogenous objects in the image. The details of the algorithm are omitted here, but can be found in [21].

4   Experimental Results

The proposed method has been tested and compared with conventional methods. This was done by performing experiments on a well-known benchmark face database, namely, the AT&T database The face database captioned "AT&T", formerly the ORL database of faces, consists of ten different images of 40 distinct subjects for a total of 400 images. Each subject is positioned upright in front of a dark homogeneous background. The size of each image is 112×92 pixels for a total dimensionality of 10304.

To construct the dissimilarity matrix, we first selected all training samples as the representative set. Then, we measured the dissimilarities between each sample and the prototypes with the dissimilarity measuring methods, such as the Hausdorff Distance (HD), the spatially Weighted Hausdorff Distance (WHD), the Gray-level Hausdorff Distance (GHD), and the spatially Weighted Gray-level Hausdorff Distance (WGHD). Since the construction of the dissimilarity matrix is a very time-consuming job, in this experiment, we constructed a 50×50 dissimilarity matrix, instead of 400×400 matrix, after selecting five objects from the AT&T database [7].

---

[7] We experimented with the simpler 50×50 matrix here. However, it should be mentioned that we can have numerous solutions, depending on the representative selection, the dissimilarity measure, and the design of classifiers in the dissimilarity space. Especially, regarding the dissimilarity measure, it is possible to construct the dissimilarity matrix rapidly by employing a computational technique. From this perspective, we are currently investigating how the experiment with the full 400×400 matrix can be performed at a high speed.

In the HD and WHD experiments, an input gray image was translated into a binary edge image by invoking an algorithm, in which edges were found by looking for "zero crossings" after filtering the image. Also, in the WHD and WGHD experiments, the weighting function of Eq. (5), $w(x)$, was defined as $1; 0.5; 0$; by using two thresholds obtained from the *mean face*, which is achieved by simply averaging all the training images.

In this paper, all experiments were performed using the "leave-one-out" strategy. To classify an image of object, that image was removed from the training set, and the dissimilarity matrix was computed with the $n-1$ images. Following this, all of the n images in the training set and the test object were translated into a dissimilarity space using the dissimilarity matrix, and recognition was performed based on the Nearest Neighbor (NN) rule. We repeated this n times for every sample and obtained a final result by averaging them.

In order to investigate the advantage gained by utilizing the proposed method, the following experiments have been conducted: First of all, the fifty facial images of *five* individuals were randomly selected from AT&T database. Next, for the facial images, the classification performance was evaluated with the HD, WHD, GHD, and WGHD methods. After repeating the above two steps ten times, the performances were finally averaged.

Table 1 shows a comparison of the averaged classification accuracy rates (%) and the averaged processing CPU-times (seconds) for the fifty facial images. Here, the fifty images have been translated into binary images for HD and WHD methods. For GHD and WGHD methods, however, the ordinary 256 gray-level input images have been used without any translation. Also, $k = 8$ was employed as the number of gray-levels to be referred when computing the distance in (9).

From Table 1, it should be noted that it is possible to improve the performance of DBCs by effectively measuring the dissimilarity. This improvement can be seen by observing how the classification accuracy rates (%) and the processing CPU-times (seconds) change. The results from Table 1 demonstrate that the classification accuracies of WHD and WGHD are improved from those of HD and GHD, respectively, while there is "marginally" change in the processing CPU-times. From the table, it is also clear that the classification accuracy of DBCs is the highest, while its standard deviation (σ) is the lowest when WGHD is used.

In review, among the four methods, it is not easy to crown one particular method with the superiority over the others in solving the dissimilarity

Table 1. A comparison of the averaged classification accuracy rates (%) and the averaged processing CPU-times (seconds) of the Dissimilarity-Based Classifiers (DBCs). The processing CPU-time of the second row is presented as an exponential form. For example, $3.45e3 = 3.45 \times 10^3$. Also, the numbers represented in the brackets of each row are the standard deviations. The details of the table are discussed in the text.

Database	HD	WHD	GHD	WGHD
AT&T	85.00 (5.18)	93.00 (4.82)	81.20 (10.72)	99.00 (1.05)
	3.45e3 (1.64e2)	3.02e3 (1.02e2)	5.93e4 (2.22e2)	5.92e4 (2.12e2)

measuring problem. However, as a matter of comparison, it is clear that with regard to the classification accuracies, the proposed dissimilarity measuring method, WGHD, is better than the conventional schemes.

5 Conclusions

In this paper, a method that seeks to optimize Dissimilarity-Based Classifiers (DBCs) by using a newly modified Hausdorff distance, the spatially Weighted Gray-Level Hausdorff Distance, was considered. To construct the dissimilarity matrix of DBCs, the dissimilarity was measured directly from the input gray-level image without extracting the binary edge image from it. Thus, the problems caused by the binary edge map could be overcome. Also, instead of obtaining the distance on the basis of the entire image, we employed the spatially weighted mask by which the entire image region was divided into several subregions according to their importance.

The proposed method has been tested on a well-known face database and compared with conventional methods. The experimental results demonstrated that the proposed scheme is better than conventional ones in terms of the classification accuracies. Although this paper has shown that DBCs could be optimized with the proposed Hausdorff distance, many tasks remain unchallenged. One of them is to improve the classification efficiency by designing a suitable classifier (i.e., linear or, possibly, quadratic classifier) in the dissimilarity space. The research concerning this is a future aim of the authors.

References

1. A. K. Jain, R. P. W. Duin, J. Mao.: Statistical pattern recognition: A review. *IEEE Trans. Pattern Anal. and Machine Intell.*, **PAMI-22(1)** 4–7 2000.
2. R. P. W. Duin, D. Ridder, D. M. J. Tax.: Experiments with a featureless approach to pattern recognition. *Pattern Recognition Letters*, **18** 1159–1166 1997.
3. R. P. W. Duin, E. Pekalska, D. de Ridder.: Relational discriminant analysis. *Pattern Recognition Letters*, **20** 1175–1181 1999.
4. E. Pekalska, R. P. W. Duin.: Dissimilarity representations allow for buiilding good classifiers. *Pattern Recognition Letters*, **23** 943–956 2002.
5. E. Pekalska.: Dissimilarity representations in pattern recognition. Concepts, theory and applications. *Ph.D. thesis, Delft University of Technology, Delft, The Netherlands*, 2005.
6. Y. Horikawa.: On properties of nearest neighbor classifiers for high-dimensional patterns in dissimilarity-based classification. *IEICE Trans. Information & Systems*, **J88-D-II(4)** 813–817 2005.
7. E. Pekalska, R. P. W. Duin, P. Paclik.: Prototype selection for dissimilarity-based classifiers. *Pattern Recognition*, **39** 189–208 2006.
8. R. P. W. Duin.: Personal communication.
9. S. -W. Kim, B. J. Oommen.: On optimizing dissimilarity-based classification using prototype reduction schemes. *This paper will be presented at the ICIAR-2006, the 2006 International Conference on Image Analysis and Recognition, in Povoa de Varzim, Portugal, in September 2006.*

10. S. -W. Kim.: On On using a dissimilarity representation method to solve the small sample size problem for face recognition. *This paper will be presented at the Acvis 2006, Advanced Concepts for Intelligent Vision Systems, in Antwerp, Belgium, in September 2006.*
11. P. N. Belhumeour, J. P. Hespanha, D. J. Kriegman.: Eigenfaces vs. Fisherfaces: Recognition using class specific linear projection. *IEEE Trans. Pattern Anal. and Machine Intell.*, **PAMI-19(7)** 711–720 1997.
12. H. Yu, J. Yang.: A direct LDA algorithm for high-dimensional data - with application to face recognition. *Pattern Recognition*, **34** 2067–2070 2001.
13. P. Howland, J. Wang, H. Park.: Solving the small sample size problem in face reognition using generalized discriminant analysis. *Pattern Recognition*, **39** 277–287 2006.
14. J. C. Bezdek, L. I. Kuncheva.: Nearest prototype classifier designs: An experimental study. *International Journal of Intelligent Systems*, **16(12)** 1445–11473 2001.
15. B. V. Dasarathy.: *Nearest Neighbor (NN) Norms: NN Pattern Classification Techniques*. IEEE Computer Society Press, Los Alamitos, 1991.
16. S. -W. Kim, B. J. Oommen.: Enhancing prototype reduction schemes with LVQ3-type algorithms. *Pattern Recognition*, **36** 1083–1093 2003.
17. S. -W. Kim, B. J. Oommen.: Enhancing prototype reduction schemes with recursion : A method applicable for "large" data sets. *IEEE Trans. Systems, Man, and Cybernetics - Part B*, **SMC-34(3)** 1384–1397 2004.
18. D. P. Huttenlocher, G. A. Klanderman, W. J. Rucklidge.: Comparing images using the Hausdorff distance. *IEEE Trans. Pattern Anal. and Machine Intell.*, **PAMI-15(9)** 850–863 1993.
19. B. Guo, K. -M. Lam, K. -H. Lin, W. -C. Siu.: Human face recognition based on spatially weighted Hausdorff distance. *Pattern Recognition Letters*, **24** 499–507 2003.
20. C. Zhao, W. Shi, Y. Deng.: A new Hausdorff distance for image matching. *Pattern Recognition Letters*, **26** 581–586 2005.
21. B. -F. Wu, Y. -L. Chen, C. -C. Chiu.: A discriminant analysis based recursive automatic thresholding approach for image segmentation. *IEICE Trans. Inf. & Syst.*, **E88-D(7)** 1716–1723 2005.

A New Model for Classifying DNA Code Inspired by Neural Networks and FSA

Byeong Kang, Andrei Kelarev, Arthur Sale, and Ray Williams

School of Computing
University of Tasmania
Private Bag 100, Hobart
Tasmania 7001, Australia
{BHKang, Andrei.Kelarev, Arthur.Sale, R.Williams}@utas.edu.au
www.comp.utas.edu.au/users/{bhkang/,kelarev/,ahjs/,rwilliams/}

Abstract. This paper introduces a new model of classifiers $CL(V, E, \ell, r)$ designed for classifying DNA sequences and combining the flexibility of neural networks and the generality of finite state automata. Our careful and thorough verification demonstrates that the classifiers $CL(V, E, \ell, r)$ are general enough and will be capable of solving all classification tasks for any given DNA dataset. We develop a minimisation algorithm for these classifiers and include several open questions which could benefit from contributions of various researchers throughout the world.

1 Introduction

Classification of data is important in data mining, see [39]. The results of this paper make the very first essential and rather non-trivial step of work on IRGS grant allocated by the University of Tasmania for the development and investigation of new Artificial Intelligence methods for classification of DNA data collected by the School of Plant Science and CRC for Sustainable Production Forestry. This is why we are mainly interested in DNA sequences, and we record all new definitions in this case. In fact, the results and concepts of this note are applicable to larger classes of problems and can be used to classify texts and documents, see for example [3], [4], [11], [21], [22], [23], [24], [27], [28], [29], [30], [32], as well as sequences in datasets of various other kinds too.

The applications of neural networks to solving numerous practical tasks have been very well known. Many useful results have been obtained with neural networks in various applied branches. For the purposes of classifying DNA sequences it is impossible to use neural networks directly processing the sequences of nucleotides. As a guide we have to look at another very well known concept of a finite state automaton (FSA) used for analysing sequences. We refer to [6], [9], [13], [14], [15], [16], [17], [19], [20], [31], [33], [34], [38] for background and some relevant recent results on the subject. The first aim of the present paper is to generalize the architecture of neural networks in order to encompass all FSA in a new concept.

A. Hoffmann et al. (Eds.): PKAW 2006, LNAI 4303, pp. 187–198, 2006.

Let us begin by introducing a new model of classifiers $CL(V, E, \ell, r)$ as a simultaneous generalisation of neural networks and finite state automata. This model combines the flexibility of neural networks and the generality of finite state automata. It is likely that this new notion will attract the attention of researchers. We develop a minimisation algorithm for the classifiers $CL(V, E, \ell, r)$. This paper includes several challenging open questions, which could benefit from contributions of many investigators throughout the world.

Before the start of experimental investigation, first of all it is important to demonstrate that the model is suitable for handling sufficiently general classes of problems and can avoid pitfalls. The main result of this paper provides the readers with a thorough verification of the fact that the classifiers $CL(V, E, \ell, r)$ are capable of handling all classification problems for DNA sequences.

Our formal model is also related to the more general concept of a labeled graph. Labeled graphs have valuable applications in various areas and have been investigated by many researchers too. We refer to [7] for a dynamic survey on graph labeling available online from the Electronic Journal of Combinatorics (see also, for instance, [36] and [37]).

The notion of classifiers $CL(V, E, \ell, r)$ has been carefully chosen from the very beginning to combine the generality of finite state automata and the flexibility of neural networks. Our main theorem shows that the classifiers $CL(V, E, \ell, r)$ can handle all classification problems for DNA datasets given sufficient computing time. A separate section develops a minimisation algorithm for these classifiers.

Background information and preliminaries are included in Section 2. Our new model is defined in Section 3. Section 4 contains the main theorem. Several major differences between classifiers $CL(V, E, \ell, r)$, neural networks, and finite state automata are pointed out in Section 5. A minimization algorithm for classifiers $CL(V, E, \ell, r)$ is presented in Section 6. Open questions are collected in Section 7.

2   Preliminaries

We use standard concepts concerning graphs and algorithms, following [2] and [35]. Throughout the word 'graph' will mean a directed graph, which is allowed to have multiple edges and loops. We refer to [8] for preliminaries on algorithms for computational analysis of DNA sequences.

Let us refer to the monographs Baldi and Brunak [1], Durbin, Eddy, Krogh and Mitchison [5], Jones and Pevzner [10] and Mount [26] for preliminaries on bioinformatics. Here we briefly recall that every DNA molecule is a double helix consisting of two strands. Each strand is a sequence of 4 nucleotides or bases: A (adenine), C (cytosine), G (guanine), and T (thymine). According to the Watson-Crick complementarity each nucleotide in one strand is crosslinked to a complementary nucleotide in another strand, and together they form a base pair. For example, the human genome contains about 3 billion base pairs and about 35,000 genes. In each DNA molecule, A and T always complement each other: A in one strand is linked to T in the second spiral. Similarly, C and G complement each other: C in one spiral is always linked to G in another strand.

If we know one sequence, it's easy to determine its complement. Therefore the sequence of base pairs in every DNA molecule can be represented with just one string over the alphabet of four letters A,C,G,T. In this paper we consider the problem of classifying strings of letters over the alphabet

$$X = \{A, C, G, T\}.$$

Accordingly, the set of all DNA sequences is precisely the set X^* of all strings over X.

3   Main Notion

A *classifier* $CL(V, E, \ell, r)$ is a quadruple

$$CL(V, E, \ell, r) = (V, E, \ell, r), \tag{1}$$

where $V = \{v_1, \ldots, v_n\}$ is the set of vertices and E is the set of edges of a graph $G = (V, E)$ with multiple edges allowed and with each edge e labeled by a letter $\ell(e)$ of the alphabet X and a real number $r(e)$. In other words, there are two functions

$$\ell : E \to X \text{ and } r : E \to \mathbb{R}. \tag{2}$$

The *state* (or *current state*) of the classifier $CL(V, E, \ell, r)$ is a labeling of all vertices by real numbers, i.e., a function

$$s : V \to \mathbb{R}. \tag{3}$$

Notice that our model has some similarities with the concept of a finite state automaton and that of a neural network, but is different from them.

   The classifiers $CL(V, E, \ell, r)$ potentially can be used for both classification and clustering. A classification of any given set of DNA sequences is a partition of these sequences into several classes. Classifiers obtain classifications via various algorithms for supervised learning. In this way the classification is known for the given set of data. The problem is to construct a classifier that will produce this classification, so that it can then be used to determine class membership of new sequences. Initial partition is usually communicated by a supervisor to a machine learning process constructing the classifier. A different problem is that of clustering data. It deals with dividing a set of given sequences into classes not known initially, but determined according to certain measures of similarities between sequences. This is usually accomplished via a process of unsupervised learning, see [39].

   Now suppose that we want to use the classifier $CL(V, E, \ell, r)$ to analyse a DNA sequence

$$x_1, x_2, \ldots, x_N, \tag{4}$$

where $x_1, \ldots, x_N \in X$. The initial state $s_0 : V \to \mathbb{R}$ can be chosen arbitrarily depending on practical implementation. Then we use the labeled graph to recursively process all letters of the sequence (4) and modify the state of the graph.

Suppose that after we have considered the first $i \geq 0$ letters of (4) the state of the graph is

$$s_i : V \to \mathbb{R}.$$

Then we can determine the next state s_{i+1} with recursion

$$s_{i+1}(v) = \sum_{w \in V, (w,v) \in E} r((w,v)) s_i(w). \qquad (5)$$

After the whole sequence (4) has been processed, for every vertex $v \in V$, we know the final value $s_N(v) \in \mathbb{R}$.

Let us now define the standard partitions which we are going to use in classification of DNA sequences. The following standard partitions will be associated with the classifier $\mathrm{CL}(V, E, \ell, r)$. For every $1 \leq k \leq N$, we define the classification $\mathcal{K}_k$ as the one which divides all given DNA sequences into classes $C_1, \ldots, C_k$, by including the sequence (4) into the class $C_i = C_i^{(k)}$, where i is chosen so that $1 \leq i \leq k$, and

$$s_N(v_i) = \max\{s_N(v_1), \ldots, s_N(v_k)\}.$$

Obviously, for $k > 1$, every classification $\mathcal{K}_k$ can be obtained from $\mathcal{K}_{k-1}$ by selecting certain elements in all classes

$$C_1^{(k-1)}, C_2^{(k-1)}, \ldots, C_{k-1}^{(k-1)}$$

of $\mathcal{K}_{k-1}$ and including them in the new class $C_k^{(k)}$. Thus, every previous classification can be regarded as a simplified version of the next one, and every next classification is a refinement of the preceding one.

4   Main Result and Verification

The main theorem of this paper establishes that the classifiers $\mathrm{CL}(V, E, \ell, r)$ are capable of solving all classification tasks for any given dataset of DNA sequences.

Theorem 1. *For each set S of DNA sequences and every given partition*

$$S = S_1 \dot{\cup} S_2 \dot{\cup} \cdots \dot{\cup} S_k \qquad (6)$$

one can find a classifier $\mathrm{CL}(V, E, \ell, r)$

$$C = (V, E, \ell, r) \qquad (7)$$

which produces classification

$$\mathcal{K} : X^* = C_1 \dot{\cup} C_2 \dot{\cup} \cdots \dot{\cup} C_k \qquad (8)$$

such that the classes of partition (6) are determined by the classes of classification (8) so that $S_i = S \cap C_i$ for all $i = 1, \ldots, k$.

Proof. First, let us define convenient notation which will enable us to refer to all sequences and their base pairs. Putting $N = |S|$, denote the sequences of the set S by $b^{(1)}, b^{(2)}, \ldots, b^{(N)}$. For each $i = 1, \ldots, N$, denote the bases of the sequence $b^{(i)}$ by the symbols $b_j^{(i)}$, where $j = 1, \ldots, m_i$ so that

$$b^{(i)} = b_1^{(i)}, b_2^{(i)}, \ldots, b_{m_i}^{(i)},$$

for all $i = 1, \ldots, N$. Suppose that the sequence $b^{(i)}$ belongs to the class $S_{\phi(i)}$ of partition (6), where ϕ is a function from $[1 : N]$ into $[1 : k]$.

Next, we introduce the following sets of vertices for the classifier $\mathrm{CL}(V, E, \ell, r)$ we are going to construct:

$$V_0 = \{v_1, v_2, \ldots, v_k\} \tag{9}$$

$$V_i = \{v_1^{(i)}, v_2^{(i)}, \ldots, v_{m_i-1}^{(i)}\} \tag{10}$$

for $i = 1, 2, \ldots, N$. In addition, choose a vertex v_0 which does not belong to any of the sets $V_0, V_1, \ldots, V_N$ and suppose that these sets are pairwise disjoint and all of their vertices are distinct. Put

$$V = V_0 \cup V_1 \cup \cdots \cup V_N \cup \{v_0\}. \tag{11}$$

To simplify further notation and have uniform definitions, we are going to denote one and the same vertex v_0 by several alternative symbols $v_0^{(1)}, v_0^{(2)}, \ldots, v_0^{(N)}$ too. Similarly, for $i = 1, 2, \ldots, N$, we introduce a new symbol $v_{m_i}^{(i)}$ to be used as an alternative notation for the vertex $v_{\phi(i)} \in V_0$. For $i = 1, 2, \ldots, N$, let us introduce sets of edges

$$E_i = \{(v_0, v_1^{(i)}) = (v_0^{(i)}, v_1^{(i)}), (v_1^{(i)}, v_2^{(i)}), \ldots, (v_{m_i-1}^{(i)}, v_{m_i}^{(i)}) = (v_{m_i}^{(i)}, v_{\phi(i)})\} \tag{12}$$

and put

$$E = E_1 \cup E_2 \cup \ldots \cup E_N. \tag{13}$$

It remains to define the initial state s_0 and the labels ℓ and r, see (2) and (3). For all $i = 1, 2, \ldots, N$ and $j = 1, 2, \ldots, m_i$, put

$$\ell((v_{j-1}^{(i)}, v_j^{(i)})) = b_j^{(i)}, \tag{14}$$

$$r((v_{j-1}^{(i)}, v_j^{(i)})) = 1. \tag{15}$$

The initial state s_0 is defined by putting, for $v \in V$,

$$s_0(v) = \begin{cases} 1 & \text{if } v = v_0, \\ 0 & \text{otherwise.} \end{cases} \tag{16}$$

This completes the definition of the classifier $\mathrm{CL}(V, E, \ell, r)$.

Suppose that the classifier is used to process the sequence $b^{(i)}$, where $1 \leq i \leq N$. We are going to show by induction that after considering the first j bases of the sequence the current state of the classifier will satisfy

$$s_j(v) = \begin{cases} 1 & \text{if } v = v_j^{(i)}, \\ 0 & \text{otherwise,} \end{cases} \tag{17}$$

for any $v \in V$.

The induction basis is provided by (16). Suppose that equalities (17) have been established for some $1 < j < m_i$. Then we can find the next state $s_{j+1}(v)$ using recursion (5).

First, consider the case where $v = v_{j+1}^{(i)}$. Since E contains only one edge of the form $(w, v_{j+1}^{(i)})$, and $s_j(v_j^{(i)}) = 1$ by the induction assumption, (15) and (5) yield us that

$$s_{j+1}(v) = \sum_{w \in V, (w,v) \in E} r((w,v)) s_j(w) \tag{18}$$

$$= \sum_{w \in V, (w,v_{j+1}^{(i)}) \in E} r((w, v_{j+1}^{(i)})) s_j(w) \tag{19}$$

$$= r((v_j^{(i)}, v_{j+1}^{(i)})) s_j(v_j^{(i)}) \tag{20}$$

$$= s_j(v_j^{(i)}) \tag{21}$$

$$= 1. \tag{22}$$

Thus, for $v = v_{j+1}^{(i)}$, the required version of (17) holds indeed.

Second, assume that $v \neq v_{j+1}^{(i)}$. Consider any $w \in V$. If $w = v_j^{(i)}$, then $(w, v) \notin E$ by the choice of v. If, however, $w \neq v_j^{(i)}$, then $s_j(w) = 0$ by the induction assumption. Thus all summands in recursion (5) vanish and we get

$$s_{j+1}(v) = \sum_{w \in V, (w,v) \in E} r((w,v)) s_j(w) \tag{23}$$

$$= 0. \tag{24}$$

This means that the desired version of (17) holds if $v \neq v_{j+1}^{(i)}$, too. By the principle of mathematical induction, it follows that (17) is always satisfied.

After all bases $b_1^{(i)}, \ldots, b_{m_i}^{(i)}$ of the sequence $b^{(i)}$ have been processed, the final state of the the classifier $CL(V, E, \ell, r)$ turns into

$$s_{m_i}(v) = \begin{cases} 1 & \text{if } v = v_{m_i}^{(i)} = v_{\phi(i)}, \\ 0 & \text{otherwise.} \end{cases} \tag{25}$$

According to our definition $b^{(i)}$ belongs to the class $C_{\phi(i)}$ of the classification $\mathcal{K}_k$, as required. This means that our classifier indeed produces a classification that agrees with the given partition of data, and so the proof is complete.

5   Neural Networks and Finite State Automata

Let us begin by comparing the classifiers $CL(V, E, \ell, r)$ with neural networks. Neural networks can be represented with similar labeled graphs. In this case the vertices are called neurons, and the labels of the edges are called weights. Edge labels are modified while a neural network is being trained. After that during the operation of the network the labels remain unchanged. Each neuron of the network takes a weighted sum of its inputs and passes it through a threshold function, usually the sigmoid function. As indicated above, the classifiers $CL(V, E, \ell, r)$ are different from neural networks and finite state automata.

The major difference is that neural networks and classifiers $CL(V, E, \ell, r)$ are designed to solve substantially different types of problems. Neural networks cannot be directly applied to classification of DNA sequences without collections of some additional data, for example, from microarrays. The reason for this is that the operation of every neural network depends on a relatively small number of input parameters, represented as continuous real values. Small changes to the values of these parameters are not generally supposed to create changes to the classification outcome. Hence it is impossible to encode whole long DNA sequences in this way. In contrast, the classifiers $CL(V, E, \ell, r)$ can process all base pairs of a given DNA sequence in succession.

Sophisticated continuous threshold functions used in neural networks lead to another serious difference (see [25], Section 11). Although the current state of a classifier $CL(V, E, \ell, r)$ appears similar to the state of a neural network, the transition to the next state is accomplished in a completely different fashion.

Comparing the classifiers $CL(V, E, \ell, r)$ to finite state automata, let us just note that each finite state automaton is used to divide its input into two classes only. Besides, the edges of finite state automata do not have real numbers as labels. These labels are inspired by analogy with neural networks. They make classifiers $CL(V, E, \ell, r)$ more flexible than finite state automata. This is why it is natural to expect that future research will demonstrate the possibility of substantial reduction to the size of the classifiers $CL(V, E, \ell, r)$ designed to handle certain classification tasks.

6   Main Algorithm

After a classifier $CL(V, E, \ell, r)$ has been found, the next natural step is to make it smaller. This can be achieved by identifying equivalent vertices. We say that a classifier $CL(V, E, \ell, r)$ is *minimal* if it can no longer be simplified by combining and identifying its vertices in some groups. As a guide to developing our minimization algorithm we are going to use the established standard terminology for analogous situations known in automata theory. Our new algorithm originates from the reduction algorithm for finite state automata described in several books (see, for example, [13], Section 3.7).

The minimization algorithm we are going to develop applies only to classifiers of the special type used in the proof of our main theorem. Namely, here we restrict our attention to the classifiers where each current state is a characteristic function of one of the vertices: it is equal to 1 at this vertex, and is equal to 0 at all other vertices. The special vertex will be called the *vertex* of the current state.

The algorithm proceeds by identifying equivalent vertices, so that one can combine them without affecting the action of the classifier $CL(V, E, \ell, r)$ on input strings.

The concepts of equivalence and congruence will be used in order to simplify the classifiers $CL(V, E, \ell, r)$. They formalise and provide exact meaning to the idea of dividing all vertices of a the classifier $CL(V, E, \ell, r)$ into groups in such a way that the operation of the classifier remains unchanged if all vertices in each group are regarded as one new vertex.

Consider a classifier $C = (V, E, \ell, r)$. In order to define the concept of a congruence on C, we begin with a few auxiliary notions. First of all, let us recall the definition of an equivalence relation. It is required, because every partition of the set of vertices V into classes can be achieved using an equivalence relation. Every subset of the set

$$V \times V = \{(u, v) \mid u, v \in V\} \tag{26}$$

is called a *relation* on the set V of all vertices. A relation ϱ is said to be *symmetric* if $(u, v) \in \varrho$ implies $(v, u) \in \varrho$, for all $u, v \in V$. It is *transitive* if $(u, v), (v, w) \in \varrho$ implies $(u, w) \in \varrho$, for all $u, v, w \in V$. A relation is said to be *reflexive* if it contains the set

$$\{(v, v) \mid v \in V\}. \tag{27}$$

An *equivalence* relation is a relation which is reflexive, symmetric and transitive.

If ϱ is a relation on V and $v \in V$, then we put

$$v^\varrho = \{w \mid (v, w) \in \varrho\}. \tag{28}$$

The set v^ϱ is called the equivalence class of ϱ containing v. It is known and easy to verify that ϱ is an equivalence relation if and only if the sets v^ϱ, $v \in V$, form a partition of V into several equivalence classes.

Let ϱ be an equivalence relation on C. Next, we show how ϱ simplifies C by combining all vertices which belong to the same classes of ϱ. The resulting classifier will be called a quotient classifier. Namely, the *quotient classifier* C/ϱ is the quadruple

$$C/\varrho = (V/\varrho, E/\varrho, \ell/\varrho, r/\varrho), \tag{29}$$

where the sets V/ϱ, E/ϱ and functions ℓ/ϱ, r/ϱ are defined as follows. The set V/ϱ is the set of all equivalence classes of ϱ on V. The set E/ϱ will contains an edge (u^ϱ, v^ϱ) with

$$(\ell/\varrho)((u^\varrho, v^\varrho)) = x \in X \tag{30}$$

if and only if there exist $u' \in u^\varrho$ and $v' \in v^\varrho$ such that $(u', v') \in E$ and $\ell((u', v')) = x$. In this case we set

$$(r/\varrho)((u^\varrho, v^\varrho)) = \sum_{u' \in u^\varrho, v' \in v^\varrho, (u',v') \in E, \ell((u',v'))=x} r((u', v')). \tag{31}$$

To simplify notation, we will use the same symbols ℓ and r for the functions ℓ/ϱ and r/ϱ, too.

We say that two vertices of a the classifier $CL(V, E, \ell, r)$ are **-equivalent* if the result of classification of each word by the classifier $CL(V, E, \ell, r)$ starting in the state of one of vertices coincides with its classification result when it starts from the state of the second vertex.

In order to determine whether two vertices are *-equivalent, the algorithm uses an iterative process based on k-equivalence. Two states are said to be k-*equivalent* if every word of length $\leq k$ produces identical classification outcomes in the case where the classifier $CL(V, E, \ell, r)$ starts in the state of the first vertex, exactly as when it starts in the second vertex. It is straightforward to verify that *-equivalence is a congruence.

In order to start the process, let us say that two vertices s and t of the the classifier $CL(V, E, \ell, r)$ $C = (V, E, \ell, r)$ are 0-*equivalent* to each other if and only if they coincide. Next, suppose that for some $k \geq 0$ the k-equivalence has already been defined. Taking any two vertices s and t in V, we say that s is $(k + 1)$-equivalent to t if and only if s and t are k-equivalent and, for each input letter $x \in X$, if the classifier starts in the state of the vertex s and processes the letter x, then it arrives at exactly the same state that is achieved if it starts in the state of the vertex t and processes the letter x from that state, so that there is no difference between starting from s or from t.

The method of computing the k-equivalence classes from $(k - 1)$-equivalence classes is a dynamic programming algorithm. It finds the k-equivalence classes by subdividing the $(k - 1)$-equivalence classes according to the change of state of the classifier $CL(V, E, \ell, r)$ when it reads each of the letters in X.

Since the set of all vertices is finite, they cannot be combined indefinitely, and at some stage the algorithm terminates. For some integer $k \geq 0$, the set of k-equivalence classes will coincide with the set of $(k + 1)$-equivalence classes. At this stage we see that both k-equivalence and $(k + 1)$-equivalence are in fact the *-equivalence.

These explanations show that the following steps find a minimal classifier $CL(V, E, \ell, r)$ equivalent to the original one:

1. Find the set of 0-equivalence classes of V.

2. For $k = 0, 1, 2$, and so on, if k-equivalence classes have been found, then divide them as described above to find the $(k + 1)$-equivalence classes of V. Stop when the set of $(k + 1)$-equivalence classes is equal to the set of k-equivalence classes, for some integer k. This step gives the set of *-equivalence classes, as explained above.

3. Construct the minimal classifier $CL(V, E, \ell, r)$ by identifying all vertices of the classes of *-equivalence.

7 Open Questions

Problem 1. Develop a minimization algorithm for classifiers $CL(V, E, \ell, r)$ with arbitrary current state functions.

Problem 2. Evaluate the running time and develop more efficient minimization algorithms for these classifiers.

Problem 3. Develop more robust algorithms by introducing a preprocessing step which will augment the dataset with other sequences and achieve similar classifications for sequences which are similar.

Two other related models used in the analysis of DNA sequences are Markov Models and probabilistic finite state automata, see Baldi and Brunak [1], Durbin, Eddy, Krogh and Mitchison [5], Jones and Pevzner [10] and Mount [26]. They have been used to identify and classify segments of one DNA sequence and are different from our model. It may make sense to explore the possibility of using these notions to classify sets of whole large DNA sequences too. This leads to the following questions suggested by the referees of this paper.

Problem 4. Investigate the running times and compare the classifications produced by our new model with those which can be obtained using Markov Models.

Problem 5. Investigate the running times and compare the classifications produced by our new model with those which can be obtained using probabilistic finite state automata.

Acknowledgements

This research has been supported by the IRGS grant K14313 of the University of Tasmania and Discovery grant DP0449469 from the Australian Research Council.

The authors are grateful to the referees for suggesting interesting open questions recorded in Probems 4 and 5.

References

1. Baldi, P. and Brunak, S.: "Bioinformatics : The Machine Learning Approach". Cambridge, Mass, MIT Press, (2001).
2. Cormen, T.H., Leiserson, C.E., Rivest, R.L. and Stein, C.: "Introduction to Algorithms", The MIT Press, Cambridge, 2001.
3. Dazeley, R.P., Kang, B.H.: Weighted MCRDR: deriving information about relationships between classifications in MCRDR, AI 2003: Advances in Artificial Intelligence, Perth, Australia, 2003, 245–255.
4. Dazeley, R.P., Kang, B.H.: An online classification and prediction hybrid system for knowledge discovery in databases, Proc. AISAT 2004, The 2nd Internat. Conf. Artificial Intelligence in Science and Technology, Hobart, Tasmania, 2004, 114–119.
5. Durbin, R., Eddy, S.R., Krogh, A. and Mitchison, G.: "Biological Sequence Analysis". Cambridge University Press (1999).
6. Eilenberg, S.: "Automata, Languages, and Machines". Vol. A,B, Academic Press, New York, 1974.
7. Gallian, J.A.: Graph labeling, Electronic J. Combinatorics, Dynamic Survey DS6, January 20, 2005, 148pp, www.combinatorics.org
8. Gusfield, D.: "Algorithms on Strings, Trees, and Sequences", Computer Science and Computational Biology, Cambridge University Press, Cambridge, 1997.

9. Holub, J., Iliopoulos, C.S., Melichar, B. Mouchard, L.: Distributed string matching using finite automata, "Combinatorial Algorithms". AWOCA 99, Perth, 114–127.
10. Jones, N.C. and Pevzner, P.A.: An Introduction to Bioinformatics Algorithms. Cambridge, Mass, MIT Press, (2004). http://www.bioalgorithms.info/
11. Kang, B.H.: "Pacific Knowledge Acquisition Workshop". Auckland, New Zealand, 2004.
12. Kelarev, A.V.: "Ring Constructions and Applications". World Scientific, 2002.
13. Kelarev, A.V.: "Graph Algebras and Automata". Marcel Dekker, 2003.
14. Kelarev, A.V., Miller, M. and Sokratova, O.V.: Directed graphs and closure properties for languages. "Proc.12 Australasian Workshop on Combinatorial Algorithms" (Ed. E.T. Baskoro), Putri Gunung Hotel, Lembang, Bandung, Indonesia, July 14–17, 2001, 118–125.
15. Kelarev, A.V., Miller, M. and Sokratova, O.V.: Languages recognized by two-sided automata of graphs. Proc. Estonian Akademy of Science **54** (2005) (1), 46–54.
16. Kelarev, A.V. and Sokratova, O.V.: Languages recognized by a class of finite automata. Acta Cybernetica **15** (2001), 45–52.
17. Kelarev, A.V. and Sokratova, O.V.: Directed graphs and syntactic algebras of tree languages. J. Automata, Languages & Combinatorics **6** (2001)(3), 305–311.
18. Kelarev, A.V. and Sokratova, O.V.: Two algorithms for languages recognized by graph algebras. Internat. J. Computer Math. **79** (2002)(12) 1317–1327.
19. Kelarev, A.V. and Sokratova, O.V.: On congruences of automata defined by directed graphs. Theoret. Computer Science **301** (2003), 31–43.
20. Kelarev, A.V. and Trotter, P.G.: A combinatorial property of automata, languages and their syntactic monoids. Proceedings of the Internat. Conf. Words, Languages and Combinatorics III, Kyoto, Japan, 2003, 228–239.
21. Lee, K.H., Kay, J., Kang, B.H.: Keyword association network: a statistical multi-term indexing approach for document categorization. Proc. Fifth Australasian Document Computing Symposium, Brisbane, Australia, (2000) 9 - 16.
22. Lee, K., Kay, J., Kang, B.H.: KAN and RinSCut: lazy linear classifier and rank-in-score threshold in similarity-based text categorization. Proc. ICML-2002 Workshop on Text Learning, University of New South Wales, Sydney, Australia , 36-43 (2002)
23. Lee, K.H., Kay, J., Kang, B.H., Rosebrock, U.: A comparative study on statistical machine learning algorithms and thresholding strategies for automatic text categorization. Proc. PRICAI 2002, Tokyo, Japan, (2002) 444–453.
24. Lee, K.H., Kang, B.H.: A new framework for uncertainty sampling: exploiting uncertain and positive-certain examples in similarity-based text classification. Proc. Internat. Conf. on Information Technology: Coding and Computing (ITCC2004), Las Vegas, Nevada, 2004, 12pp.
25. Luger, G.F, "Artificial Intelligence. Structures and Strategies for Complex Problem Solving". Addison-Wesley, 2005.
26. Mount, D.: "Bioinformatics: Sequence and Genome Analysis". Cold Spring Harbor Laboratory, (2001). http://www.bioinformaticsonline.org/
27. Park, S.S., Kim, Y., Park, G., Kang, B.H., Compton, P.: Automated information mediator for HTML and XML Based Web information delivery service. Proc. 18th Australian Joint Conf. on Artificial Intelligence , Sydney, 2005, 401–404.
28. Park, G.S., Kim, Y.S., Kang, B.H.: Synamic mobile content adaptation according to various delivery contexts. J. Security Engineering 2 (2005) 202-208.
29. Park, G.S., Kim, Y.T., Kim, Y., Kang, B.H.: SOAP message processing performance enhancement by simplifying system architecture. J. Security Engineering 2 (2005) 163-170.

30. Park, G.S., Park, S., Kim, Y., Kang, B.H.: Intelligent web document classification using incrementally changing training data Set, J. Security Engineering 2 (2005) 186–191.
31. Păun, G. and Salomaa, A.: "New Trends in Formal Languages". Springer-Verlag, Berlin, 1997.
32. Petrovskiy, M.: Probability estimation in error correcting output coding framework using game theory. AI 2005: Advances in Artificial Intelligence, Sydney, Australia, 2005, Lect. Notes Artificial Intelligence **3809** (2005) 186–196.
33. Pin, J.E.: "Formal Properties of Finite Automata and Applications". Lect. Notes Computer Science **386**, Springer, New York, 1989.
34. Rozenberg, G. and Salomaa, A.: "Handbook of Formal Languages". Vol. 1, Word, Language, Grammar, Springer-Verlag, Berlin, 1997.
35. Smyth, B.: "Computing Patterns in Strings". Addison-Wesley, 2003.
36. Sugeng, K.A., Miller, M., Slamin and Bača, M.: (a, d)-edge-antimagic total labelings of caterpillars. Lecture Notes in Comput. Sci. **3330** (2005) 169–180.
37. Tuga, M. and Miller, M.: Δ-optimum exclusive sum labeling of certain graphs with radius one. Lecture Notes in Comput. Sci. **3330** (2005) 216–225.
38. van Leeuwen, J.: "Handbook of Theoretical Computer Science". Vol. A,B, Algorithms and Complexity. Elsevier, Amsterdam, 1990.
39. Witten, I.H. and Frank, E.: "Data Mining: Practical Machine Learning Tools and Techniques with Java Implementations". Morgan Kaufmann, 2005.

Improvements on Common Vector Approach Using k-Clustering Method

Seohoon Jin[1], MyungWoo Nam[2], and Sang-Tae Han[3]

[1] Department of Cross Sell Marketing, Hyundai Capital, 15-21, Youido-Dong, Youngdungpo-Gu, Seoul, 150-706, Korea
[2] Department of Digital Elecronics Design, Hyejeon College, San 16, Namjang-Ri, Hongsung-Eup, Hongsung-Gun, Choongnam, 350-702, Korea
[3] Department of Informational Statistics, Hoseo University, 29-1, Asan, 336-795, Korea
sthan@office.hoseo.ac.kr

Abstract. In this paper, an advanced common vector approach (CVA) method for isolated word recognition is presented. The proposed method eliminates drawback of conventional CVA method, which is impossibility of being applied to a large number of training voices case, by dividing the training voices into a few small groups where those voices belong to a class of one of the spoken words. The results from using MFCC, LPC, LSP, Cepstrum, and auditory model shows that the proposed method solves the drawback of conventional CVA method. It got better recognition rate of 1.39% without significant changes of amounts of computation.

1 Introduction

Voice signal contains psychological and physiological properties of speakers as well as dialect differences, acoustical environment effects, and phase differences [1]. Because of these reasons, even the same voice signal shows different characteristics when the sound comes from a different speaker. These characteristics of voice signal make it difficult to extract the common properties from the voice class(word or phoneme). Most of the speech recognition methods recognize the voice with the following process; extract the common properties of the word, make comparative patterns of the observed voice, compare both of them. Therefore, the efficiency of recognition method is totally depends on both extraction of the common property and decision of words using comparison. In this paper, we proposed advanced CVA method. The algorithm of CVA is easy to extract the common properties from training voice signals and also does not need complex calculation [1-4]. In addition, CVA has shown high accuracy in recognition results. CVA, however, has a problem for applying when many training voices are given [4]. Generally, to get the optimal common vectors from one of voice classes, various voices should be used for training. However it is impossible to get continuous high accuracy in recognition with CVA because CVA has a limitation to use many training voices. To solve the problem and improve recognition rate, k-clustering method is used. Various experiments were performed using voice

A. Hoffmann et al. (Eds.): PKAW 2006, LNAI 4303, pp. 199–206, 2006.

signal database made by ETRI to prove the validity of proposed method. The result of experiments shows improvements in performance. The problem of CVA can be solved without calculation problem using the proposed method.

2   Common Vector Approach

Let us use the following notations.

R^n : n-dimensional vector space,
$\langle x, y \rangle$: scalar product of vector x and y,
$||x||$: Euclidean norm of vector x.

Let there be given n-dimensional linearly independent vectors $a_1, a_2, \ldots, a_m \in R^n$ for $m < n$. Here each a_i ($i = 1, 2, \cdots, m$ denotes the speaker number) belongs to the class of one of the spoken words and it will be used as the training set. Each a_i can be written as summation of a common vector x, which represents the common properties of the class of the spoken words and the difference vector $a_{i,diff}$, which represents the properties of the i_{th} person.

$$
\begin{aligned}
a_1 &= x + a_{1,diff} \\
a_2 &= x + a_{2,diff} \\
&\vdots \\
a_m &= x + a_{m,diff}
\end{aligned}
\tag{1}
$$

Let us define,

$$
\begin{aligned}
b_1 &= a_1 - a_m \\
b_2 &= a_2 - a_m \\
&\vdots \\
b_{m-1} &= a_{m-1} - a_m.
\end{aligned}
\tag{2}
$$

From the linear algebra, it is well known that the orthonormal vector set $\{z_1, z_2, \cdots, z_{m-1}\}$ with property $\langle z_i, z_j \rangle = \{$ 1 if $i = j$; 0 if $i \neq j$ $\}$ can be obtained from this basis $\{b_1, b_2, \cdots, b_{m-1}\}$ by using Gram-Schmid orthogonalization method. Let

$$
\bar{a}_i = \langle a_i, z_1 \rangle z_1 + \langle a_i, z_2 \rangle z_2 + \cdots + \langle a_i, z_{m-1} \rangle z_{m-1}, (i = 1, 2, \ldots, m).
\tag{3}
$$

Each $\bar{a}_i$ denotes the summation of the projections of a_i on the orthonormal base of the difference subspace B. The vector $\tilde{a}_i = a_i - \bar{a}_i$ is called the common vector $\tilde{a}_{common}$ for the set of vector $a_1, a_2, \ldots, a_m$ that belongs to the same word class.

$$
\tilde{a}_{common} = \tilde{a}_i = a_i - \bar{a}_i, (i = 1, 2, \ldots, m)
\tag{4}
$$

3   Common Vectors Using k-Clustering Method

The k-clustering method permits common vector extraction regardless of the number of training voice. In addition to that, it can train various voice signals without significant increase of amounts of computation. The k-clustering method consists of three parts greatly. First part is the k-clustering which clusters training voice signals by small subspace of k, next part is the common vector extraction which extracts common vector from small subspaces, and the last part derives the total common vector from extracted common vectors of subspaces. We explain whole algorithm of the k-clustering method and verify a detailed numerical expressions. Let there be given n-dimensional linearly independent vectors $a_1, a_2, \ldots, a_m \in R^n$ for $m \gg n$. Here each $a_i (i = 1, 2, \ldots, m$ denotes the speaker number) belongs to the class of one of the spoken words. First, cluster vector a_i into small subspace of k, (where $k < m$ and $k \leq n - 1$). Next, extract the common features using conventional CVA from small subspace of k. The orthonormal vector set of the subgroup is used for extraction of each common vector. By the last process, extract the total common vector from common vectors of k subgroups again. Extracted total common vector means the common component of vector a_i, and we can get 100% recognition rate for the training word set.

3.1 k-Clustering Method

There exist various methods that cluster vector a_i into small subspace of k. In this paper, we consider two methods of clustering. The first method clusters vectors at random, and the other method uses k-means algorithm [5-6]. From the results of experiment, the case of clustering vectors at random and using k-means algorithm could not get the invariable recognition rate. Along with how to cluster vector a_i, each small subspace is changed, and it effects to the recognition rate. Clustering methods do not make small subspace as the same form in always. In k-means algorithm, elements of small subspace are changed according to how to establish elementary value. Therefore, if we form small subspace using above methods, we might get the good recognition rate but the bad recognition rate as well. It can be a big problem when we apply them to practical recognizer. In this paper we used the best clustering result that was gotten from several independent trials. The used method of clustering training vectors by using k-means is as following.

Goal. Cluster vector $a_i \in R^n$, $(i = 1, 2, \ldots, m)$ by small subspace of k where, $k < m$ and $k \leq n - 1$.

Step 1. Get a reference vector r from a_i .
Step 2. Divide the remainings except r by k and form small subspaces.

```
Begin initialize m, k, u₁, u₂, ..., uₖ
        do classify m − 1 samples into k groups according to nearest uᵢ
            recompute uᵢ (which is the cluster mean)
        until no change in uᵢ
        return u₁, u₂, ..., uₖ
    end
```

Step 3. Include r to each small subspace.

3.2   Finding First Common Vectors

The method that extracts common vector from each small subspace is equal with the conventional CVA where the reference vector r should be used. The CVA in j_{th} small subspace can be written as following. Suppose a_i^j ($j \leq k, i = 1, 2, \ldots, t < n - 1$) is a vector of j_{th} small subspace. From Eq. (5), the common vector of j_{th} small subspace can be written as Eq. (6).

$$
\begin{aligned}
b_1 &= a_1^j - r \\
b_2 &= a_2^j - r \\
&\vdots \\
b_t &= a_t^j - r
\end{aligned}
\tag{5}
$$

$$
\tilde{a}_i{}^j = \tilde{a}_{common}^j = a_i^j - \overline{a}_i^j = r - \overline{r}, (i = 1, 2, \ldots, t)
\tag{6}
$$

$$
\tilde{a}_i^j = \langle a_i^j, z_1^j \rangle z_1^j + \langle a_i^j, z_2^j \rangle z_2^j + \cdots + \langle a_i^j, z_t^j \rangle z_t^j
\tag{7}
$$

3.3   Finding Total Common Vector

The common vector $\tilde{a}_{common}^j$ extracted from each small subspace can be written Eq. (8) and it is linearly independent.

$$
\begin{aligned}
\tilde{a}_{common}^1 &= \tilde{a}_{common}^{total} + \tilde{a}_{common,diff}^1 \\
\tilde{a}_{common}^2 &= \tilde{a}_{common}^{total} + \tilde{a}_{common,diff}^2 \\
&\vdots \\
\tilde{a}_{common}^k &= \tilde{a}_{common}^{total} + \tilde{a}_{common,diff}^k
\end{aligned}
\tag{8}
$$

From Eq. (8), $\tilde{a}_{common}^{total}$ means the total common vector and $\tilde{a}_{common,diff}^j$ represents individual features of the common vector extracted from j_{th} small subspace. Therefore, we can get the common component of total training vectors by extracting the common vector from first common vectors using CVA once more. The derived total common vector becomes a reference vector that is used to decide a word in the decision part. If the input vector enters, firstly, project the input vector into the each small subspace and derive the difference between the input vector and the projected vector. Secondly, calculate the each likelihood of the common vector of small subspaces. Lastly, compare likelihood with the total common vector by projecting the common vector of small subspace that has the maximum likelihood into the subspace of the total common vector again. Using Fig.1 process, we decide the recognition word that has the most likelihood finally. The proposed method has the similar calculation amount to the conventional method and it solves the problem of the limitation of the number of training vector.

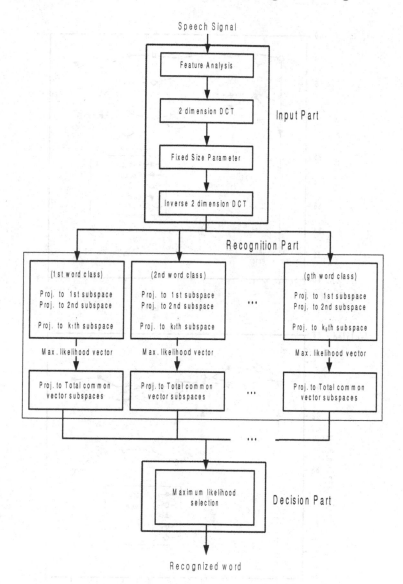

Fig. 1. Block diagram of k-clustering method

4 Experiments and Results

The ETRI isolated word database, which consists of the 20 digits (0-9) and 22 words, was used to evaluate the performance of the proposed algorithm. In the 20 digits, 20 male and 20 female speakers each recorded 4 repetitions of each word, for a total of 3200 utterances. In the 22 words, 48 male and 43 of female of speakers each recorded 1 repetition of each word, for a total of 2002 utterances.

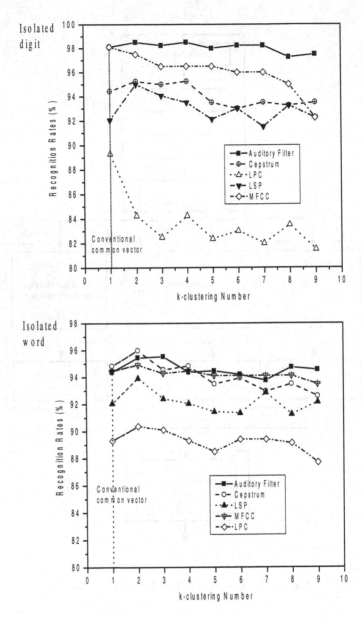

Fig. 2. Recognition rates of variations of the k-clustering number for isolated digit and isolated word

The input speech is sampled at 8 kHz rate and stored by 16 bits. The feature is extracted every 15 msec frame overlapped by 50%. To confirm effectiveness of proposed method, we made an experiment using auditory model (32_{nd} order),

Cepstrum (32_{nd} order), LPC (12_{th} order), LSP (32_{nd} order) and MFCC (32_{nd} order) [7-9]. The extracted parameter is normalized using two-dimensional DCT in fixed size and used input of CVA and k-clustering method [10]. From the results, we find that the change of the number of small subspace do not cause big effect to recognition result, but the recognition rate slightly decreased along with increasing the number of clusters.

Table 1. Average recognition rates(%) of isolated digit by number of clusters

Method	Conventional CVA($k=1$)	$k=2$	$k=4$	$k=6$	$k=8$
MFCC	98.12	97.50	96.50	96.00	95.00
Auditory Model	98.12	98.50	98.50	98.25	97.25
Cepstrum	94.43	95.25	95.25	93.00	93.25

Table 2. Average recognition rates(%) of isolated word by number of clusters

Method	Conventional CVA($k=1$)	$k=2$	$k=4$	$k=6$	$k=8$
MFCC	94.91	94.47	94.11	94.11	93.49
Auditory Model	94.38	95.45	94.38	94.20	94.74
Cepstrum	94.38	95.98	94.83	93.93	93.49

5   Conclusion

The algorithm of CVA is easy to extract the common properties from training voice signals and does not need complex calculation. The CVA has shown high accuracy in the recognition results. However the CVA has a drawback of being impossible to use for many training voices. In this paper, we proposed the k-clustering method which improved the CVA, and experimented Korean speaker independent isolated word recognition. The k-clustering method solved the drawback of CVA and got better recognition rate of 1.39% without significant changes of amounts of computation. Proposed method, however, has various recognition rate according to the number of clusters. Therefore, determination of the optimal number of clusters will be critical for applying k-clustering CVA. In this study the number of clusters are explored heuristically but several criteria can be applied and compared for finding optimal number of clusters. If the optimal number of cluster problem is solved the algorithm of k-clustering CVA will be simpler to implement. There will be further research for developing an algorithm to find isolated word recognition with various clustering algorithms such as fuzzy c-means clustering.

References

1. Bilginer, M. et al.: A novel approach to isolated word recognition, Speech and Audio Processing, IEEE Trans. **7** (1999) 620-628
2. Cevikalp, Hakan, Wilkes, Mitch: Discriminative common vectors for face recognition, IEEE Trans. Pattern analysis and machine intelligence **27**, no.1, Jan. (2005)
3. Gulmezoglu, M. B., Dzhafarov, V. and Barkana, A.: The common vector approach and its relation to the principal component analysis, IEEE Trans. Speech and Audio Processing **9** (2001) 655-662
4. Gulmezoglu, M. B., Dzhafarov, V. and Barkana, A.: Comparison of common vector approach and other subspace methods in case of sufficient data, in Proc. 8th conference on Signal Processing and Applications, Belek, Turkey (2000) 13-18
5. Duda, Richard O., Hart, Peter E., Stork, David G.: Pattern Classification, Wiley Interscience (2001)
6. Cho, C.H.: Modified k-means algorithm, The Journal of Acoustical Society of Korea **19** no.7 (2000) 23-27.
7. Wallace, G. K.: The JPEG still picture compression standard, Consumer Electron., IEEE Trans. **38** (1992) 18-34.
8. Lay, David C.: Linear algebra and its applications, Addison Wesley (2000)
9. Deller, John R., Proakis, Jhon G., Hansen, John H. L.: Discrete-time processing of speech signals, Macmillan Publishing Company. (1993)
10. Nam, M.W., Park, K.H., Jeong, S.G., Rho, S.Y.: Fast algorithm for recognition of Korean isolated word, The Journal of Acoustical Society of Korea **20**, no.1 (2001) 50-55

The Method for the Unknown Word Classification

Hyunjang Kong[1], Myunggwon Hwang[1], and Pankoo Kim[2,*]

[1] Dept. of Computer Engineering, Chosun University,
375 Seosuk-dong Dong-Ku Gwangju 501-759 South Korea
{kisofire, mghwang}@chosun.ac.kr
[2] Dept. of Computer Engineering, Chosun University
pkkim@chosun.ac.kr

Abstract. Natural Language Processing is a hard task. For the real Natural Language Processing, it is a necessary technique to process the unknown words. In this paper, we introduce the method for understanding the unknown words means. Many terms are newly created and we do not find these words in dictionary. Unknown words are generally occurred by reflecting the new phenomenon and technology. Hence, unknown words are dramatically created because of rapid changes in society. However, it is a hard task to define the meaning of all unknown words in dictionary. So, in this paper, we focus on how the machine understands the unknown words means. We propose a method to classify unknown words using the relevancy values between all nouns in the document and their TF values.

1 Introduction

For the real Natural Language Processing(NLP), there are several tasks that we have to solve. One of the significant tasks is to classify the unknown words. The NLP system encounters words that are not in its lexicon frequently. In here, we define the terms as the unknown words which do not exist in the lexicon. As a NLP system will perform well, it should understand the unknown words. Even when the unknown words are not occurred very often, they have an effect on a NLP system quality. For the past several years, the importance of the task for Natural Language Processing systems has been recognized. However, it is a very hard task because the huge unknown words can be created everyday and they cannot be completely registered in the lexicon by the human. Therefore, robust approaches for processing unknown words are needed. [1][2][3] The method presented in this paper allows the automatic detection of the unknown words means. A machine applying our technique can understand the unknown words and it can classify the words automatically instead of the human knowledge engineer. In our approach, several algorithms such as unknown words detection, unknown words understanding and classification are demanded. Our proposed method is using the *Relevancy* values among the terms based on WordNet and their *TF* values.

This paper organized as follows: Section 2 introduces the background technologies and we present our proposed method for understanding the unknown words

---

* Corresponding author.

A. Hoffmann et al. (Eds.): PKAW 2006, LNAI 4303, pp. 207–215, 2006.

automatically. Section 3 presents the experimental results and evaluation. Finally, we conclude our study in Section 4.

2 Background Knowledge and Our Approach

In this section we describe our approach how to understand the unknown words means. Under this view, unknown word processing is based on the existing techniques.

a) *TF* values
b) *Relevancy* values in WordNet
c) Noun detection

The brief explanation about each technique is as follows.

In this paper, we use WordNet to detect the nouns in the sentences and to get the relevancy values among these nouns. WordNet is a semantic lexicon for the English language. It groups English words into sets of synonyms called synsets and records the various semantic relations between these synonym sets. And English nouns, verbs, adjectives and adverbs are organized into synonym sets in WordNet.[4] Most synsets are connected to other synsets via semantic relations. In our approach, we consider the nouns and two semantic relations (*Hypernym/Hyponym, Holonym/Meronym*) in Word-Net for measuring the relevancy values. Second basic technique in our approach is the TF value. Documents are written using huge terms. In this case, it is very hard to determine which words are most important in documents. Until now, for determining the importance of words in document, we generally measure the *TF*(term frequency).[5] And the formula calculating *TF* is as follows:

$$tf = \frac{n_i}{\sum_k n_k}$$

In above formula, $\sum_k n_k$ means the total of all terms, which the document contains, n_i is the number of occurrence of the specific term. And values gained through *TF* formula are the indispensable data with the relations between terms in our approach.

Processing unknown words is a mixture of two above techniques. In order to process document containing unknown words, we use all the nouns contained in the document. In our approach, unknown words are processed following the steps.

STEP 1. Parsing the document to detect unknown word
Unknown words are frequently created because of new theories, technologies, change of the customs and human activities and many manufactures. As we briefly mentioned before, we define unknown words that are not published in the lexical as WordNet. Principle of unknown word acquisition in our approach is as follows:

(1) Detect nouns in document.
(2) Search the unknown words based on pseudo code in table 1. In here, we use the WordNet as the standard lexicon to detect the unknown word.
(3) According to the results of the pseudo code, we are able to detect unknown words.

Table 1. The pseudo code for detecting the unknown word

> C_i : Set of nouns ;
> WN : Nouns in WordNet ;
> If $(C_i \ulcorner\in WN)$
> {
>   Unknown_Word = C_i ;
> }
> Unknown_Word : The results after processing STEP 1

STEP 2. Extracting the sentences containing unknown word and all nouns except unknown word in document

In this step, we extract the sentences containing unknown words in document. And then, we detect the all nouns in document except unknown word. In our method, we assume that all nouns in document are related to the unknown word. Moreover, we think the nouns in sentences, which contain unknown word, are closely related to unknown word. Hence, we divide the document into two parts:

- Sentences Part: All sentences containing the unknown word
- Nouns Part: All nouns in document except the unknown word

The processing flow of sentences extraction is as follows:

(1) Use the unknown word detected through STEP 1.
(2) Extract the sentences, which include the unknown word based on pseudo code in Table 2.
(3) Through the STEP 2, we can prepare the input data for our method. For the input data, we analyze the document as we divide into the two parts like Table 3. For example, in this processing, we try to define the meaning of 'S-Class', which is an unknown word.

Table 2. The pseudo code for extracting the sentences containing unknown word

> Unknown_Word : The result through STEP 1
> S_i : All sentences in document
> If $(Unknown\_Word \in S_i)$
> {
>   {S} = {S} + S_i ;
> }
> {S} : The result after processing STEP 2

Table 3. Extracted sentences containing new word and extracted nouns

Original contents of document
Origin of S-Class
While the exact beginning of the colloquial S-Class expression ... it always referred to the most spacio us and largest luxury vehicle ... applied to vehicles requiring premium fuel or "Super" due to higher co mpression ratio and output of the top-of-the-line engines.
Widely distributed, over time, these cars came to be considered desirable ...
The S-Class grew out of the modest "Ponton" model, a six cylinder sedan known ...
The line was introduced with the 220a, 219 (W105), 220S, and 220SE sedan, coupe, ...
... the W110 featuring a shorter hood and wheelbase for the "economy" models 190c and 190Dc, and t he 300 SE (W112]]), a short time predecessor ...
... saw the opportunity to build much larger vehicles, including the limited volume 1964 600 limousine ...
... sedan (still W111) along the new larger (W108/W109).
These larger vehicles established the S-Class reputation that continues ...
... brought aerodynamics to the previously brick shaped cars.
The W140 saw the car grow dramatically in its proportions ...

Extracted Sentences containing unknown word	All Nouns in document
Origin of S-Class	Car, sedan, accelerator, gear, seat,
While the exact beginning of the colloquial S-Class expression	break, system, auto, vehicle, carria
... it always referred to the most spacious and largest luxury veh	ge, airbag, space, car, car, person,
icle ... applied to vehicles requiring premium fuel or "Super" du	wheel, sedan, vehicle, passenger, n
e to higher compression ratio and output of the top-of-the-line e	ews, window, coupe, accident, car
ngines.	...
The S-Class grew out of the modest "Ponton" model, a six cylind	
er sedan known ...	*(in here, we detect all nouns and d*
These larger vehicles established the S-Class reputation that co	*o not consider the duplication abo*
ntinues ...	*ut nouns)*

STEP 3. Measure the relevancy values and TF values

In STEP 2, we detected the nouns in the document as shown in Table 3. In our approach, we consider the *TF* values of all extracted nouns and the relevancy values among the nouns based on the relations in WordNet. STEP 3 is processed as follows:

(1) Calculate the *TF* values based on pseudo code in Table 4.

Table 4. The pseudo code for calculating the *TF* values

```
N : The nouns in nouns part
TFV : TF values
For (N_i ∈ Nouns Part)
{
        TFV_i = TFV_i + 1;
}
TFV_i : The result after processing TF calculation
```

Using pseudo code in Table 4, *TF* values about all nouns in table 3, for example, are calculated as shown in Table 5.

Table 5. Calculated TF values about sample document as shown in table 3

Synset_ID	TF Value	Words
04207742	11	System
02853224	8	Car
10464998	8	Model
05327145	7	Gear
14401902	3	Sedan
04008331	1	......

Table 5 shows the results of the *TF* values of all nouns in Nouns part. Until now, in previous approaches to understand the unknown words, the *TF* values play the core role. However, it does not support for the complete understanding of unknown words. Hence, we consider the *Relevancy* value between nouns as well as *TF* values for the perfect unknown words understanding.

(2) Measure the *Relevancy* values between noun in Sentence part and noun in Nouns part based on pseudo code in Table 6.

Table 6. The pseudo code for measuring the *Relevancy* value

```
N : The nouns in Nouns part
SN : The nouns in Sentences part
RV : Relevancy Value
For (N_i ∈ N) {
    If (N_i Related to SN)
    {
            RV_i = RV_i + 1;
    }
}
RV_i : The result after measuring the Relevancy value
```

After detecting the nouns, we measure the *Relevancy* value, which means how the nouns are related to the unknown word. In here, we use the semantic relations between nouns defined in WordNet. As we mentioned before, there are many relations in WordNet. However, we just use two relations (*Hypernym/Hyponym, Holonym/Meronym*) for measuring the *Relevancy* value among nouns.

Table 7. Measured *Relevancy* values among nouns about sample document as shown in table 3

Synset_ID	Related Synset_ID and relationship based on WordNet	Relevancy Values
04207742	~ 10720570 ~ 09729204 %p 04410590 %p 04254824 ~ 04008331 %p 03400842 %p 03389509 %p 03228252 ~ 03006338	9
02853224	~ 10066029 ~ 09933701 ~ 09908263 ~ 09843239 ~ 09843239 ~ 09819657 ~ 09721227 ~ 09624379 ~ 09608190 ~ 09598437 ~ 09463859 ~ 09385835 ~ 09285577 ~ 09223355 ~ 09145707 ~ 09144663 ~ 09019701 ~ 09013278 ~ 09012224 %p 04916889 #m 07463651 @ 00003226	22
08103697	~ 08134364 ~ 08124727 ~ 08120943 ~ 08087842 ~ 08066770 ~ 08063710 ~ 08034339 ~ 07992043 ~ 07982095	9
00017572	~ 14254673 ~ 14130483 ~ 14051444 ~ 14051242 ~ 14050897 ~ 14027638	6

Table 7 shows the measured *Relevancy* values among nouns based on the relations defined in WordNet.

STEP 4. Defining the unknown words mean
STEP 4 is the core in our approach. STEP 4 exploiting the *Relevancy* values and *TF* values of the nouns is processed as follows:

(1) Use the *TF* value and *Relevancy* value each noun.
(2) Figure out the meaning of the unknown word as we calculate *CV* values about each noun based on pseudo code in Table 8.

Table 8. The pseudo code for measuring the *Relevancy* value

```
N : The nouns in Nouns part
TF : TF Value
RV : Relevancy Value
CV : Concept Value
For (Nᵢ ∈ N)
{
        N_TFᵢ : TF Value of Ni
        N_RVᵢ : Relevancy Value of Ni
        CVᵢ = N_RVᵢ * N_TFᵢ ;
}
CVᵢ : The result after calculating the Concept Value
```

(3) Through the STEP 4, we can determine the meaning of the unknown word. Using the results in Table 9, the unknown word(S-Class) could be classified into the car.

Table 9. CV Results

Nouns	Results	
(Synset_ID)	CV	Concepts
02853224	**798.0001**	**Car, auto, automobile, machine, motorcar**
05394410	352.7999	Arrangement, organization, organization, system
05396456	338.8001	Design, plan
04008331	313.6001	Sedan
07924048	293.9999	System, scheme

Table 9 shows the *CV* results about each noun. In table 9, noun 'car' has the highest *CV* values. So, we could conclude that 'S-Class' is very close to the car. Hence, efficiency of our approach was certified correct.

3   Experimental Results and Evaluation

In example of Section 2, we are sure that our approach is suitable for defining the meaning of unknown words. We have evaluated our proposed method formatively. In order to examine the validity of the method we adopted, the approach was evaluated with some amounts of document resources. Our testing environment is as follows:

(1) Objective: To measure efficiency of our approach.
(2) Materials: WordNet as knowledge source, each 5 documents containing five unknown words – TGV(train), Fanta(beverage), Zindane(soccer player), S-Class(car) and Marlboro(cigarette).
(3) Results: Table 10 shows the results of *CV* values of the five documents respectively.

Table 10. Experimental results

Document 1	CV	Document 2	CV	Document 3	CV	Document 4	CV	Document 5	CV
TGV									
train	8	France	7	France	4	road	24	train	2
Korea	8	travel	6	train	2	train	18	station	2
Pusan	8	pass	4	transportation	2	way	15	Paris	2
route	7	rail	4	rail	2	track	12	Sur	2
South	7	train	4	passenger	2	car	8	rail	2
Fanta									
can	6	Thailand	81	food	48	orange	8	orange	144
mango	5	mango	13	orange	9	company	4	lemon	12
orange	4	product	12	soda	8	war	4	can	6
diet	3	drink	11	product	8	coca	4	flavor	4
lemon	3	orange	11	flavor	7	Germany	2	taste	4
Zindane									
ball	10	France	70	man	12	soccer	6	game	2
cup	5	Paris	30	world	11	world	6	football	2
player	5	world	11	player	7	cup	4	generation	2
name	4	cup	8	cup	7	France	4	trailer	2
French	4	Frenchman	7	game	6	player	3	player	2
S-Class									
sedan	6	car	15	car	24	car	24	sedan	13
car	5	symbol	8	system	18	model	9	price	10
equipment	3	road	8	body	12	system	7	research	8
vehicle	3	coupe	6	time	12	seat	7	car	7
road	2	system	6	control	11	body	4	model	3
Marlboro									
history	14	cigarette	8	cigarette	3	cigarette	35	television	30
advertising	12	brand	5	brand	3	search	9	Cancer	8
life	10	man	3	sales	2	image	7	death	8
marketing	6	Morris	3	world	2	brand	6	man	7
cigarette	5	Philip	3	country	2	man	5	cigarette	4

In the results of the testing, about the tangible unknown words such as *TGV, Fanta, S-Class* and *Marlboro*, the effective classification is possible using our method. However, the results of the testing about the notional unknown words such as *Zindane* have lower accuracy than the tangible words. Based on the testing results, we determine the meaning of each unknown word. Figure 1 shows the classification of the unknown words means using the formula (1).

$$Unknown\ Words\ Means(= Highest\ Value(C_i)),\quad C_i = \sum W(N_i) \qquad (1)$$

where, W is the weight value and N is the noun in the result. In formula (1), we give the W value (*5, 4, 3, 2, 1*) respectively to the each noun in each document by sequence of the *CV* values. And then, we measure the final result using formula (1). Finally, we assume that the word, which has the highest value, is most related to the unknown word. Hence, we classify the unknown words means into the word, which is selected through formula (1).

Through the experimental results in Figure 1, we expect the efficient unknown word classification using our approach. Especially, the results about *Malboro(means Cigarette), TGV(means Train) and S-Class(means Car)* are very satisfied for supporting the performance of our approach. Hence, we certified that the robust unknown word classification is possible using our approach.

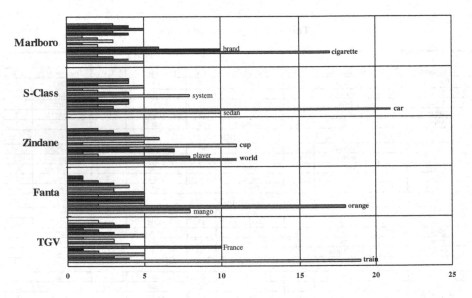

Fig. 1. Experimental Results

4 Conclusion

In this paper, we try to understand the meaning of the unknown words. Nowadays the unknown words are dramatically increasing and study on processing unknown words has been not researched although this task is very important for real Natural Language Processing. In our approach, we assume that nouns in document are related to the unknown word. Therefore, we calculated the *CV* values using the *TF* values and *Relevancy* values of the nouns. Based on the testing results, we assure that the *CV* values strongly reflect the fact which noun is most related to the unknown word.

Acknowledgement

This research was supported by the Program for the Training of Graduate Students in Regional Innovation which was conducted by the Ministry of Commerce Industry and Energy of the Korean Government.

References

1. H. Ishikawa, A. Ito and S. Makino, 1993. Unknown Word Processing Using Bunsetsu-automaton, 2nd class of Technical report of IEICE, LK-92-17, pp.1-8 (in Japanese).
2. T. Kamioka and Y. Anzai, 1988. Syntactic Analysis of Sentences with unknown words by Abduction Mechanism, Journal of Artificial Intelligence, Vol.3, pp.627-638 (in Japanese).
3. C. Kubomura, T. Sakurai and H. Kameda, 1996. Evaluation of Algorithms for Unknown Word Acquisition, Technical report of IEICE, TL96-6, pp.21-30 (in Japanese).
4. Scott, S., Matwin, S.: Text Classification using WordNet Hypernyms. In the Proceeding of Workshop – Usage of WordNet in Natural Language Processing Systems, Montreal, Canada (1998).
5. Gelbukh, A., Sidorov, G., Guzman, A.: Use of a Weighted Topic Hierarchy for Document Classification. In Václav Matoušek et al (eds.): Text, Speech and Dialogue in Poc. 2nd International Workshop. Lecture Notes in Artificial Intelligence, No.92, ISBN 3-540- 66494-7, Springer-Verlag., Czech Republic (1999) 130-135.

An Ontology Supported Approach to Learn Term to Concept Mapping

Valentina Ceausu and Sylvie Desprès

Paris 5 University,
Paris, 75006,
France
ceausu@math-info.univ-paris5.fr, sd@math-info.univ-paris5.fr

Abstract. We propose in this paper an approach to learn term to concept mapping with the joint utilization of an existing ontology and verb relations. This is a non-supervised solution that can be applied to any field for which an ontology modeling verbs as relations holding between the concepts was already created. Conceptual graphs are learned from a natural language corpus by using part-of-speech information and statistic measures. Labeling strategies are proposed to assign terms of the corpus to concepts of the ontology by taking into account the structure of the ontology and the extracted conceptual graphs. This paper presents the approach proposed to learn the conceptual graphs from the corpus and the labeling strategies. A first experimentation in the field of accidentology was done and its results are also presented.

Keywords: concept learning, ontology, verb relation.

1 Introduction

The rapid evolution in the production of documents in natural language requires to define efficient automated approaches allowing to find relevant information in those documents. This paper presents an approach that uses verb relations and a domain ontology to assign terms of a given corpus to concepts of the field. Those assignations can be used thereafter for various exploitation scenarios, that is to say: indexing collection of documents, annotating documents, etc. This is un unsupervised approach, unless the use of a domain ontology to support the process. The task to achieve could be described as fallows : let O be a domain ontology and C a collection of domain-specific texts. Our goal is to identify within C terms W representing linguistic expression of concepts of O ontology.

Three steps have been proposed to carry out this labelling process: In a first stage, verb relations are extracted from the corpus. Each verb relation represents a triple composed of a verb and a pair of terms connected by this verb. In a second phase, statistical processing is performed to structure those verb relations as conceptual graphs. The last phase is based on the assumption that

A. Hoffmann et al. (Eds.): PKAW 2006, LNAI 4303, pp. 216–222, 2006.

the field ontology models verbs of the field as relations holding between the concepts. If this is the case, labelling strategies are using the ontology and extracted conceptual graphs to assign field specific terms to field concepts.

We shall approach that topic by answering a number of questions: which method should be used to extract verb relations from corpus? How to learn conceptual graphs from the extracted verb relations? Those questions are analyzed in sections 2 and 3. Given a domain ontology and a set on conceptual graphs, which strategies will be used to assign terms to concepts? The solution is discussed in section 4. A first experimentation in the field of accidentology is described and its results are presented in section 5. Related approach are presented in section 6. Conclusions and perspectives of this work will end the paper.

2   Extracting Verb Relations from Corpus

To extract verb relations from corpus, we adopted a pattern recognition approach. This approach is using part-of-speech information and consists in seeking within the corpus particular associations of lexical categories. Such an association represents a lexical pattern. A set of lexical patterns including, among other categories, a verb, is defined. A pattern recognition algorithm, described in [1], is using part-of-speech information to identify associations of words matching the patterns of this set. The algorithm takes as input the corpus tagged by TreeTagger(see [2]) and a set of lexical patterns including verbs. It is applied at sentence level and it automatically generates a set of word regroupings matching those patterns. Obtained word regroupings can be[1]: a verb relation, such as: *véhicule diriger vers bretelle (vehicle direct to slip road)*; an incomplete verb relation: *piéton traverser (pedestrian crossing)* or *diriger vers l'opéra (direct to opera)*; or meaningless word regroupings:*c,véhicule (C, vehicle,) ; venir de i (come from i)*.

3   Learning Conceptual Graphs

The goal of this phase is to learn conceptual graphs from the results of pattern recognition algorithm.

A conceptual graph represents a hierarchy having as a top a verb and, on a second layer, arguments connected to the verb by theirs grammatical function, subject or object. We use the term *conceptual graph* as it was introduced by [3]. A conceptual graph corresponds to a set of verb relations generated by the same verb. To learn conceptual graphs, a chain of treatments are performed that are based on lexical similarity measures presented bellow.

3.1   Lexical Similarities

A similarity measure associates a real number R to a pair of strings $S1, S2$. [4] presents a number of approaches to calculate similarities between strings.

---

[1] Examples of this paper are translated in English, although they are extracted from a French corpus experimentation.

Similarity measures implemented for this work are detailed further. Jaccard co-efficient considers a string composed of several sub-strings and calculates the similarity between two strings S and T as :

$$Jaccard(S,T) = \frac{|S \cap T|}{|S \cup T|}$$

This measure is given by the number of sub-strings common to S and T compared to the number of all sub-strings of T and S. If we consider characters as sub-strings, the coefficient expresses the similarity by taking into account the number of common characters of S and T only.

Jaro and Jaro-Winkler coefficients, introduced below, express the similarity by taking into account the number and the position of characters shared by S and T. Let $a = a_i..a_k$ and $b = b_1..b_l$ be two strings. A character $a_i \in s$ is considered common to both strings if there is a $b_j \in t$ such as: $a_i = b_j$ and $i - H \leq j \leq i + H$, where $H = \frac{min(|S|,|T|)}{2}$. Let $s^1 = a_1^1..a_k^1$ characters of s common to t and $t^1 = b_1^1..b_l^1$ characters of t common to s. We define a transposition between s^1 and b^1 as an index i such as: $a_i^1 \neq b_i^1$. If T_{s^1,t^1} is the number of transpositions from s^1 to t^1 the Jaro coefficient calculates the similarity between s and t as follows:

$$Jaro(s,t) = \frac{1}{3}(\frac{|s^1|}{|s|} + \frac{|t^1|}{|t|} + \frac{|s^1| - T_{s^1,t^1}}{|s^1|})$$

[5] proposes a version of this coefficient by using P, the length of the longer prefix common to both strings. Let $P^1 = max(P,4)$, then Jaro-Winkler is written:

$$Jaro - Winckler(s,t) = Jaro(s,t) + \frac{P^1}{10}(1 - Jaro(s,t))$$

There are also hybrid approaches calculating similarities recursively, by analyz-ing sub-strings of initial strings. Thus, Monge-Elkan uses two steps to calculate similarity between $s^1 = a_1^1..a_k^1$ and $t^1 = b_1^1..b_l^1$: the two strings are divided into sub-strings then the similarity is given by:

$$sim(s,t) = \frac{1}{k} \sum_{i=1}^{k} max_{j=1}^{L}(sim^1(a_j,b_j))$$

where values of $sim^1(a_j,b_j)$ are given by some similarity function, called basic function, for example one of those previously presented. Such a function is called a *level 2 function*. For this work, Monge-Elkan is implemented by using the coefficients of Jaccard, Jaro and Jaro-Winkler as a basic function.

Statistic measures will be used in different phases of our approach.

3.2   An Iterative Approach to Learn Conceptual Graphs

Conceptual graphs are learned from the set of word regroupings extracted ac-cording to section 2. An iterative solution is proposed, performing a number of steps, each of them adding a new layer to the graphs.

The first step identifies verb classes, that represent the set of verb relations generated by the same verb. For each verb class, instances of *Verb* and *Verb, Preposition* patterns are added to the set of roots. We argue that for verbs accepting prepositions, each *verb, preposition* structure is specific and for this reason we create conceptual graphs for any of those structures. This step creates a number of conceptual graphs having one level, which is to say the root.

For each root, its arguments are identified : terms that are subjects and objects. This step is adding a second layer to each conceptual graph.

As for a given verb, arguments can have different levels of granularity, a new level can be added to conceptual graphs by clustering those arguments.

A cluster is a group of similar terms, having a central term C called centroid and its k nearest neighbors. Based on the observation that the greater number of words in a word regrouping there are, the more specific his meaning is, an algorithm is proposed to cluster arguments of verb relations. This algorithm is considering single word arguments as centroids, and it uses the Monge-Elkan similarity coefficient to add terms to clusters.

4   Term to Concept Mapping Using the Ontology

To assing arguments of verb relations to concepts we make the assumption that, for a given conceptual graph, the verb R representing its top node is already modelled by the ontology. Let r be the corresponding relation and $Range^r$, $Domain^r$ concepts of the ontology connected by r. $Domain^r$ will be used to label subject arguments, while $Range^r$ will be used to label object arguments. Assignation of terms to concepts is performed by one of labelling strategies described bellow.

A first strategy ignores the hierarchical organization of arguments. Thus, similarities between each argument and concepts of the ontology are calculated using one of presented similarity measures. The argument is assigned to the concept maximizing this similarity if the value of similarity is greater than a pre-defined threshold, otherwise it will be labelled as *unknown*.

Two strategies are proposed which take into account the hierarchical structure of arguments. Therefore, each cluster is considered as a hierarchy having on its first level the centroid and on its second level terms that are specializations of centroid. The second strategy is a top-down strategy. In the first phase, it identifies concepts of ontology which label the centroid of the cluster. If the centroid of a cluster is labeled as unknown, the same label is assigned to each term of the cluster. If the centroid of a cluster is labeled by a concept C of ontology, labels for other terms of the cluster are searched only in the set of sub-concepts of C. In this way, the top-down labelling strategy reduces the search space.

A third strategy is based on a bottom-up approach. For each cluster, the similarities between its terms and the concepts of ontology are calculated by using one of presented coefficients. If values of similarities are higher than a threshold, the concept labels the term. If this is not the case, the term will be labeled as *unknown*. Based on the assignments of each term of cluster to ontology concepts, similarity between the centroid and a concept of ontology is given by:

$$sim(Cen, C) = \frac{1}{k} \sum_{i=1}^{k} sim(t_i, C)$$

where t_i is a term of the cluster, C is a concept of ontology, $sim(t_i, C)$ is the similarity between t_i and C and k is the number of terms labeled by C. Those three labelling strategies are used in a first experimentation in the field of accidentology which is described in the next section.

5   Experimentation in Accidentology and First Results

A first experimentation of this approach was done in the field of road accidents. It uses a corpus composed of 250 accident reports of road accidents which occurred in and around the Lille region.

We used an ontology of road accidents created from accident reports by using Terminae (see [6]). This ontology is expressed in OWL (see [7]). It models about 450 concepts describing road accidents and 300 roles connecting those concepts. Arguments that are objects of the *circuler avec (circulate with)* relation are labelled. The top-down strategy labels *véhicule blanc* as *inconnu (unknown)*. The bottom-up strategy allows us to eliminate the centroid *feu (fire)*, which is labelled as *inconnu (unknown)*. On the downside, clusters having a small number of terms are penalized.

For Jaro-Winkler coefficient, results of the three strategies are similar to results obtained with Jaro coefficient. For Jaccard coefficient, the bottom-up strategy shows a failure as it assigns the term *véhicule* to the concept *véhicule de service*. Independent of the coefficient that in used, the top-down strategy performs faster.

For the same couple *term, concept*, values of the Jaccard coefficient are slightly lower than values of Jaro and Jaro-Winkler. Therefore, values of thresholds for Jaccard coefficient need to be lower than thresholds of Jaro and Jaro-Winkler coefficients.

6   Related Work

Approaches proposed in different application fields, such as ontology learning or word-sense disambiguation are at the origin of this work. Among them, [8] propose Asium, a machine learning system which acquires subcategorization frames of verbs based on syntactic input. Asium system hierarchically clusters nouns based on the verbs that they are syntactically related with and vice versa.

[9] propose RelExt, a system which is capable of automatically identifying highly relevant triples (pairs of concepts connected by a relation). RelExt extracts relevant terms and verbs from a given text collection and it estimates

relations between them through a combination of linguistic and statistical processing.

[3] propose a system having a multi-layered architecture aiming to extract information from genetic interaction data. The system uses verb patterns modelled as conceptual sub-graphs to characterize unknown terms in sentences. The goal is to enrich an existing ontology by integrating discovered concepts.

7 Conclusion and Future Work

We have presented an approach allowing us to assign terms of a corpus to concepts of an ontology. This approach is using jointly verb relations and a domain ontology. Different measures which estimate similarity between strings have been implemented and used in the various phases of our approach.

A first experimentation in the field of road accidents shows that Jaro and Jaro-Winkler coefficients provide better similarities estimation than the Jaccard coefficient. Among the labelling strategies, the top-down strategy performs faster and generates better assignments of the terms to concepts of ontology. Those are only preliminaries conclusions and further case studies are needed.

As a further direction, a feedback could be added in order to enrich the domain ontology by integrating some of the arguments of verb relations.

References

1. Ceausu, V., Desprès, S.: Towards a text mining driven approach for terminology construction. In: 7th International conference on Terminology and Knowledge Engineering. (2005)
2. Schmid, H.: Probabilistic part-of-speech tagging using decision trees. In: International Conference on New Methods in Language Processing. (1994)
3. Roux, C., Prouxet, D., Rechenmann, F., Julliard, L.: An ontology enrichment method for a pragmatic information extraction system gathering data on genetic interactions. In: Ontology Learning Workshop at ECAI. (2000)
4. Cohen, W., Ravikumar, P., Fienberg, S.: A comparison of string distance metrics for name-matching tasks. In: IJCAI-2003,Workshop on Information Integration on the Web pages. (2003)
5. Monge, A., Elkan, C.: The field-matching problem: algorithm and applications. In: Second International Conference on Knowledge Discovery and Data Mining. (1996)
6. Biébow, B., Szulman, S.: A linguistic-based tool for the building of a domain ontology. In: International Conference on Knowledge Engineering and Knowledge Management. (1999)
7. Szulman, S., Biébow, B.: Owl et terminae. In: 14-me Journe Francophone d' Ingnierie des Connaissances. (2004)
8. Faure, D., Nedellec, C.: Asium, learning subcategorization frames and restrictions of selection. In: 10th European Conference On Machine Learning, Workshop on text mining, Chemnitz, Germany (1998)
9. Schutz, A., Buitelaar, P.: Relext: A tool for relation extraction from text in ontology extension. In: International Semantic Web Conference. (2005) 593–606

10. Alfonseca, E., Manandhar, S.: Improving an ontology refinement method with hyponymy patterns. In: Third International Conference on Language Resources and Evaluation. (2001)
11. Faatz, A., Steinmetz, R.: Ontology enrichment with texts from the www. In: SemanticWeb Mining 2nd Workshop at ECML/PKDD. (2002)
12. Monge, A., Elkan, C.: An efficient domain-independent algorithm for detecting approximately duplicate database records. In: Workshop on data mining and knowledge discovery, SIGMOD. (1997)
13. Miller, G.: Wordnet: A lexical database for english. CACM **38** (1995) 39–41

Building Corporate Knowledge Through Ontology Integration

Philip H.P. Nguyen[1] and Dan Corbett[2]

[1] Justice Technology Services, Department of Justice,
Government of South Australia, 30, Wakefield Street, Adelaide, SA 5000, Australia
nguyen.philip@saugov.sa.gov.au
[2] Science Applications International Corporation,
1710 SAIC Drive, McLean, VA 22102, USA
corbettd@saic.com

Abstract. This paper presents an approach for building corporate knowledge, defined as the total knowledge acquired by an enterprise in its business dealings, through integration of its existing ontologies. We propose to represent corporate knowledge as the final merged ontology, defined under our formalism, in which a canon, or common ontology, is used as the standard under which all other ontologies are aligned. The canon is also enriched with knowledge gained during each ontology merging exercise. Our method ensures that all resulting ontologies are semantically consistent, compact and complete, as well as mathematically sound, so that formal reasoning could be conducted.

1 Introduction

Merging businesses usually leads to consolidation of information from different sources and different classification systems. The automation of such processes has the potential to reduce costs and delays, and thus contributes to the success of the merger. However, when information has been structured on the foundation of different ontologies, full automation of the information integration process is rarely achieved. Often, domain experts and knowledge engineers are needed to assist with decision making. The usual method is to first convert or translate one ontology from its current model to the model used by the other ontology, before integrating them. To date, most tools are semi-automatic and vary according to the particular nature of the domain being treated. Tools of a more general applicability, independent from ontology domains, exist, but as a trade-off they are less automatic and rely more on experts and knowledge engineers to assist during their usage. A common note in these techniques is that they are often used by organizations that require ontology merging on an infrequent basis, usually ad-hoc or once-only. It is usually not necessary, or not possible, to transfer knowledge gained from one application to another, apart from improved skills of the people involved and improved overall methodologies. In this paper, we propose a method to merge ontologies, which also captures and accumulates knowledge gained from each application in order to facilitate subsequent

A. Hoffmann et al. (Eds.): PKAW 2006, LNAI 4303, pp. 223–229, 2006.

ontology merging exercises. This would be of particular help for large enterprises with a frequent requirement for ontology merging and enable the building of *corporate knowledge*, defined as the total knowledge acquired by an enterprise in its business dealings and which we represent under our formalism by the final merged ontology, resulting from successive mergings of the enterprise's existing ontologies.

2 Ontology Formalization and Integration

The formalization of canon and ontology in this section is a summary of [5] and could be considered as within the Conceptual Graph Theory [8][2]. However, our proposed ontology merging method is new and does not rely on the Conceptual Graph Theory.

• Canon

In simple terms, a canon is a framework for knowledge organization that models the real world through abstract concepts, relations and association rules between them.

Formally, we define a canon K as a tuple $K = (T, I, \leq, conf, B)$ where:

(1) T is the union of the set of concept types T_C and the set of relation types T_R . We assume that each of those sets is a *lattice*, in which every pair of types has a *unique* supremum and a *unique* infimum. This assumption ensures *mathematical soundness* of the structure and is common in lattice theory [10].

(2) I is the set of *individual markers*, which are real-life instances of concept types, e.g. "John" is an individual marker of the concept type "Person".

(3) $\leq$ is the subsumption relation in T, enabling definition of a hierarchy of concept types and a hierarchy of relation types.

(4) *conf* is the *conformity* relation that relates each individual marker in I to a concept type in T_C . It in effect defines the (unique) infimum of all concept types that could be used with an individual marker (called *coreferent concept nodes* in [1]), e.g. the individual marker "John" may be associated through *conf* with the concept type "Man", which is the infimum of other concept types "Person", "Mammal" and "Living Entity", and therefore "John" can be used as an instance of those concepts, i.e., "John is a man, a person, a mammal and a living entity."

(5) B is the *Canonical Basis* function that associates each relation type with an *ordered* set of concept types that may be used with that relation type, e.g., "fatherOf" is a relation type that is associated through the function B with two (identical) concept types: "Person" and "Person", in which the first "Person" is the father and the second "Person" is the child. B must also satisfy the association rule: if two relation types are related (through the subsumption relation $\leq$) then their transformations by B should also be related in the respective order.

• Ontology

We define an ontology as a semantically consistent subset of a canon, dealing with a particular domain of discourse. Conversely, a canon could be considered as a *universal* (or *upper*, or *top-level*, or *unified*, etc.) *ontology*, encompassing multiple specialist domains. For example, a classification of diseases written in English could be viewed as an ontology while a general English semantic lexicon (such as WordNet) could be regarded as a canon.

Formally, we define an ontology O on a domain M with respect to a canon $K = (T, I, \leq, conf, B)$ as a tuple $O = (T_{CO}, T_{RO}, I_O)$ where:

(1) T_{CO} (resp. T_{RO}) is a subset of T_C (resp. T_R).
(2) I_O is a subset of individual markers in I, that relate to the domain M.
(3) All other relations in K (i.e., $\leq$, $conf$ and B) and their association rules are carried over to O.

Mathematical properties of ontology as formalized above are detailed in [5].

- **Ontology Integration**

Since we define an ontology as a subset of a canon, implicitly a canon must exist before an ontology. However, in reality, in an enterprise, such a canon must be built either from scratch or from successive mergings of existing ontologies. Our paper deals with the latter. Our method ensures that each ontology merging exercise brings new knowledge to the common canon and the common canon is in turn used to make subsequent ontology merging exercises more automatic.

Our method consists of the following steps:

(1) **Step 1 - Input preparation:** In this first step, we ensure that the canon (if exists) and the source (input) ontologies are represented in a format that is suitable for our computational process, i.e., each structure simply needs to be represented as a list of supertype/subtype relationships (such as "conceptA>conceptB"). Note that in most applications, the source ontologies may be legacy systems, not strictly defined under any canon, i.e., they may contain concepts, relations and/or subsumption relations, not existing in the initial canon. For the first ontology merging exercise, we can use a semantic lexicon such as WorldNet as the initial canon.

(2) **Step 2 - Execution of the MultiOntoMerge algorithm:** The core of our method is an algorithm, called *MultiOntoMerge*, which accepts as inputs the above canon and source ontologies and produces as outputs a (draft) merged ontology and a list of concepts, relations and subsumption relations that have been identified as existing in either or both source ontologies but missing from the canon. (If the list is empty, then the process terminates as the canon already contains all the knowledge embedded in the source ontologies and the draft merged ontology is the final merged ontology). We could visualize the merging process by reasoning in a top-down manner with the lattice structures of the source ontologies. The lattices are merged *branch by branch*, starting from the top-level nodes and proceeding *depth-first*, through all their branches. To provide a common starting point, we assume, without loss of generality, that all lattices contain the common "universal" type, which is the fictitious supertype of all types. This lifts the restriction imposed in other previous work ([2][3][7]), which requires the knowledge engineer to specify an entry point for the merging process. During processing, the insertion of any new object (e.g., a new concept type) into the merged ontology is based on its compatibility with other objects already in that

structure. That compatibility is determined by the existing semantics and relations of those objects in the canon. If the new object only exists in either or both source ontologies, and does not exist in the canon initially, then it is noted in a separate list for later verification and validation (Step 3). After validation, that object will be first incorporated in the canon, then in the merged ontology (Step 4). During processing, MultiOntoMerge also performs a number of *consistency checks* on the canon, source ontologies and merged ontology, to ensure that each structure is:

> ➢ Semantically Consistent: The use of a common canon as the upper ontology for all other ontologies being treated, means that all ontologies are semantically aligned with the canon semantics.
> ➢ Semantically Compact: This means that MultiOntoMerge transforms duplicate concepts into *coreferences* (see definition in [8]), in particular when the same concept is used under different synonymous names. The algorithm also consolidates the structure by removing redundant subsumption relations, based on their transitivity property.
> ➢ Semantically Complete: This means that the algorithm ensures that if two concepts (or relations) in the structure are related (as specified by their subsumption relationship in the canon) and that relationship is not already contained in the structure, then it is added to the structure for completeness.
> ➢ Mathematically Sound: This means that the algorithm ensures that the structure is a *lattice*, as per its mathematical definition [4]. In particular, any two concepts (or any two relations) in the structure must have a *unique* supremum and a *unique* infimum, which are also contained in the structure. Note that this property is also invoked in other ontology merging techniques, such as FCA-Merge [9].

(3) **Step 3 - Validation of missing data in the canon:** The list of new objects (i.e., new concepts, relations and subsumption relations) produced by the previous step is then reviewed and validated by a domain expert. This validation is based on the semantics of the objects and is essentially manual. This expert may also consider some items on the list as errors in the source ontologies and therefore should not be carried forward into the final outputs (i.e., merged ontology and canon). These decisions are marked on the list and will be executed in Step 4.

(4) **Step 4 - Re-execution of the MultiOntoMerge algorithm:** The MultiOntoMerge algorithm described in Step 2 is re-executed with the additional input consisting of the decisions made by the expert in the previous step (in the form of a list of new objects missing from the canon, plus any possible errors in the source ontologies). The algorithm then carries out the expert decisions by correcting the source ontologies (if errors were detected there) and by incorporating the missing objects into the canon. Finally, it performs the ontology merging task as in Step 2 again. In most cases, unless further inconsistencies are detected as a result of the previous (incorrect) decisions by the expert, after this step, the output list of data missing from the canon should be empty, and the canon, the source ontologies and the merged ontology should be consistent as per our formalization (i.e. all ontologies are finally subsets of the final canon) and the processing terminates.

Notes

(1) One notable feature from our method is that, as the canon is constantly enriched after each application of MultiOntoMerge, in subsequent ontology merging exercises, missing data identified in Step 2 would be less and less significant and manual intervention in Step 3 would be less and less required. The ontology merging process therefore becomes more and more automatic after each usage.

(2) With regard to other ontology merging theories and techniques (such as Prompt, FCA-Merge, Chimaera, etc.), most of them, if not all, are semi-automatic and require interactions with knowledge engineers and experts during the construction of the merged ontology. However, unlike our method, none of them explicitly leverages knowledge gained from earlier ontology merging exercises to improve subsequent ones. On the other hand, some of them may complement our method by assisting in the automation of our "validation of missing data" step (i.e., Step 3).

(3) Our method enables discovery of new knowledge during the ontology merging process (e.g., with the identification of missing concepts and creation of new concepts in the merged ontology). This is similar to some other techniques, such as *Formal Concept Analysis* [10] and *Simple Conceptual Graph* [1] (see its "lattice-theoretic interpretation").

Example

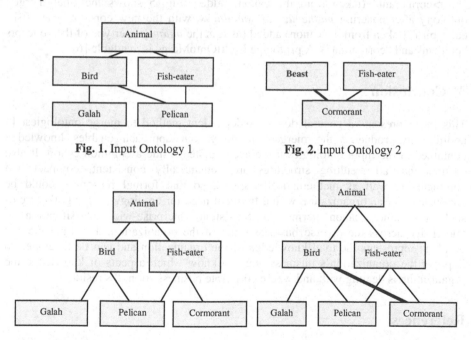

Fig. 1. Input Ontology 1

Fig. 2. Input Ontology 2

Fig. 3. Merged Ontology (after Semantic Compaction)

Fig. 4. Merged Ontology (after Semantic Completion)

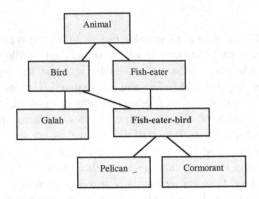

Fig. 5. Final Merged Ontology (after ensuring Mathematical Soundness)

Above is an example of merging two simple ontologies represented by Fig. 1 and 2 (with WordNet used as the initial common canon). Fig. 3 shows the merged ontology after *semantic compaction*, with "animal" and "beast" forming the same concept and the redundant relation "animal>cormorant" removed (based on the canon). Fig. 4 shows the merged ontology after *semantic completion*, with the new relation "bird>cormorant" (taken from the canon) added. Fig. 5 shows the final merged ontology after ensuring *mathematical soundness*, with the new concept type "fish-eater-bird" (taken from the canon) added (as it is the *unique* supertype of the concepts "pelican" and "cormorant"). A prototype MultiOntoMerge is available [6].

3 Conclusion

This paper presents a general domain-independent method to merge ontologies. In addition to producing the merged ontology, our approach enables knowledge contained in the input ontologies to be accumulated inside a common canon. It also ensures that all resulting structures are semantically consistent, compact and complete, as well as mathematically sound, so that formal reasoning could be conducted. For an organization with a frequent need for ontology merging, the use of such a common canon permits a consistent enterprise-wide classification of knowledge across the diverse business units of the organization. The final merged ontology represents the total knowledge of the organization and can be leveraged to improve the organization's business, e.g., to know which aspects of knowledge the organization is dealing with and where corporate business strengths reside.

References

1. Chein, M., Mugnier, M.L.: "Concept Types and Coreference in Simple Conceptual Graphs", 12th International Conference on Conceptual Structures, Huntsville, Alabama, USA (2004)
2. Corbett, D.: "Reasoning and Unification over Conceptual Graphs", Kluwer Academic Publishers, New York (2003)

3. Corbett, D.: "Filtering and Merging Knowledge Bases: a Formal Approach to Tailoring Ontologies", Revue d'Intelligence Artificielle, Special Issue on Tailored Information Delivery, Cecile Paris and Nathalie Colineau (editors) (Sept./Oct. 2004) 463 – 481
4. Ganter, B., Wille, R.: "Formal Concept Analysis: Mathematical Foundations", Springer, Heidelberg, Germany, (1996-German version) (1999-English translation)
5. Nguyen, P., Corbett, D.: "A Basic Mathematical Framework for Conceptual Graphs," IEEE Transactions on Knowledge and Data Engineering, Vol. 18, No. 2 (Feb. 2006) 261-271
6. Nguyen,P.,MultiOntoMergePrototype: http://users.on.net/~pnguyen/cgi/multiontomerge.pl
7. Nicolas, S., Moulin, B., Mineau, G.: "sesei: a CG-Based Filter for Internet Search Engines", 11th International Conference on Conceptual Structures, Dresden, Germany (2003)
8. Sowa, J.: "Knowledge Representation: Logical, Philosophical, and Computational Foundations", Brooks Cole Publishing Co., Pacific Grove, CA (1999)
9. Stumme, G., Maedche, A.: "FCA-Merge: Bottom-Up Merging of Ontologies", 7th International Conference on Artificial Intelligence, Seattle, USA (2001)
10. Wille, R.: "Restructuring Lattice Theory: an Approach based on Hierarchies of Concepts", Ordered Sets, I. Rival (ed.), Reidel, Dordrecht-Boston (1982).

Planning with Domain Rules Based on State-Independent Activation Sets[*]

Zhi-hua Jiang and Yun-fei Jiang

Dep. of Computer Science, Ji Nan University, Guang zhou, China, 510632
Software Research Institution, Zhong Shan University, Guang zhou, China, 510275
jnujzh@163.com, lncsri05@zsu.edu.cn

Abstract. In AI planning community, planning domains with derived predicates are very challenging to many planning system. Derived predicate is a new application of domain rules and domain knowledge acquisition. In this paper, we propose an approach to planning with derived predicates: defining activation sets of a derived predicate which are unrelated to any specific state and computing them in the preprocess phase through the instantiation rule-graph; replacing a derived predicate with one of its activation sets in relax-plan to extract action sequences. And we also implement the proposed approach in a new planner, called FF-DP, which shows good performance in our experiments.

Keywords: Deterministic planning; Domain rules; activation sets.

1 Introduction

Domain Rules are referred as particular domain knowledge and new knowledge is often deduced from known knowledge base by applying these rules. Domain rules, or domain axioms, are always the hotshot problem in AI planning community. Given an operator file and a problem file, an intelligent planning problem is intended to require a sequence of actions so that the original state can be transferred into the goal state by applying these actions in turn. The International Planning Competition (IPC), which is held biyearly since 1998, is referred as the most top-level academic conference and planning systems competition in this field. Derived predicate is one of two new features of PDDL2.2 language [1], the standard competition language in International Planning Competition 2004 (IPC-4). In classical planning, predicates are divided into two categories: basic and derived. While basic predicates may appear as effects of actions, derived ones may only be used in action preconditions or goal state [2]. So, derived predicates are not affected directly by domain actions, and their truth in the current state is inferred from that of basic predicates via domain axioms. PDDL2.2 introduces two benchmark domains containing derived predicates: PSR-Middle, and PROMELA (Philosophers and Optical-Telegraph) [3]. There were 19 planners that joined the classical track in IPC-4; however, only four planners (LPG-td, SGPlan, Marvin, and Downward) attempted to solve those domains containing derived predicates.

[*] This research is funded by Chinese Natural Science Foundation (60173039).

A. Hoffmann et al. (Eds.): PKAW 2006, LNAI 4303, pp. 230–237, 2006.

Therefore, the lack of the ability of dealing with derived predicates actually blocked most planners to solve more competition problems.

Some methods have been used to deal with derived predicates in different planning system; however, their feasibility and efficiency are limited. Compiling them away is firstly proposed, but is recently proved to involve a worst-case exponential blow-up in the size of the domain description or in the length of the shortest plans [2]. And Gazen and Knoblock propose a pre-processing algorithm which transforms domain axioms into equivalent 'deduce' operators, but this may lead to inefficient planning [4, 5]. Also, LPG-td planner presents an approach to planning with derived predicates where the search space consists of particular graphs of actions and rules, called rule-action graphs [5]. However, the calculation of possible activation sets of a derived predicate in a rule graph is often enormous and boring because activation sets have to be recalculated as soon as the current state changes. So we attempt to find state-independent activations sets of a derived predicate, computing them only once in preprocess phase. We implement our idea on a new planner, called FF-DP (FF-Derived Predicate), which is a modified version of FF2.3 [6].

In the rest of this paper, we first introduce the definition of state-independent activation sets of a derived predicate. Then, we discuss how to calculate state-independent activation sets in a rule-graph and use them in relax-plan. Finally, we examine FF-DP in some specific benchmark problems.

2 Rule Graphs and State-Independent Activation Sets

As we know, the truth value of a derived predicate is determined by a set of domain rules in the form: if Φ_x then $P(x)$, where $P(x)$ is the derived predicate, x is a tuple of variables, the free variables in Φ_x are exactly the variables in x, and Φ_x is a first-order formula such that the negated normal form (NNF) of Φ_x does not contain any derived predicate in negated form [1]. For example, Fig. 1 shows a classical derived predicate "above" in the blocks world: A block x is above y, if x is on y, or it is on a third block z, which is above y.

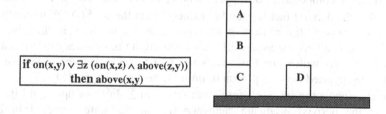

Fig. 1. Derived predicates in the blocks world

Given a rule r = (if Φ_x then $P(x)$) and a tuple of constants c ($|x| = |c|$), we can derive an equivalent set R composed of grounded rules (contain no variables) by applying the transformations mechanism (more detail in [5]). The set R can only contain basic facts or derived facts, for instance, on(A,B) and on(B,C) are basic facts, while above(A,B) and above(B,C) are derived facts. The set R that consists of grounded rules can be transformed into an AND-OR graph, which LPG-td calls "rule-graph": AND-nodes (fact-nodes) are either leaf nodes labeled by basic facts, or nodes labeled by derived facts; OR-nodes (rule-nodes) are labeled by grounded rules in R. For a rule-node, its in-edge (only one) is from the derived predicate deduced by itself, and its out-edges (often more than one) point to triggering conditions of this rule. For example, in Fig.2, the derived predicate "above(A,B)" can be inspired by rule r1,r2, or r3, and the triggering conditions of r2 are on(A,C) and above(C,B).

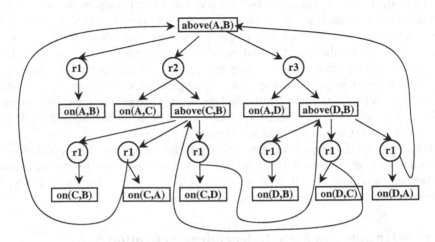

Fig. 2. A portion of the rule graph for a blocks world domain

In LPG-td, an activation set is defined as follows:" Given an unsupported derived precondition node d at a flawed level l of a RA-graph, an activation fact set for d is a minimal set F of basic facts such that $S(l) \cup F \vdash^R \Psi_d$, where Ψ_d is the derived fact represented by d." [5]. Here, S(l) means the current state. Because of the limitation of "minimal set", an already supported basic fact in the level l doesn't belong to any activation set, even if it is a triggering condition of d. Therefore, an activation set of a derived fact is composed of unsupported basic facts in the level l. The notation " $\vdash^R$" means that the derived fact Ψ_d can be deduced from the set $S(l) \cup F$ under the rule set R. So, we can see the reason of introducing activation sets is that, because that derived facts can't appear as any effects of actions but basic ones can, we may replace the derived facts with some basic facts in order to select actions to support them so that the unsupported derived precondition node becomes supported.

We can denote the set of all activation sets with Σ. For example, in Fig.2, the set Σ_{LPG} of the derived predicate "above(A,B)" in the state depicted in Fig. 1 is {{on(A,C), on(C,B)} , {on(A,C), on(C,D), on(D,B)} , {on(A,D), on(D,B)}, {on(A,D), on(D,C), on(C,B)}}. However, suppose that if we stack the block A onto the block D, the state becomes the set {table(C), table(D), on(B,C), on(A,D)} and the set Σ_{LPG} for "above(A,B)" is turned into{{on(A,B)}, {on(A,C), on(C,B)} , {on(A,C), on(C,D), on(D,B)} , {on(D,B)}, {on(D,C), on(C,B)}}. Thus an activation set defined in [5] is closely related to the current state, and once the state changes the activation sets have to be recalculated. Actually, in relax-plan of LPG-td, there exists a great many of time-consuming recalculation. Therefore, we redefine activation sets which are state-independent, as follows:

Definition 1. Given a rule-graph derived from the rule set R, a **state-independent activation set** (short for **SIAS**) for a derived predicate d is a minimal set F of basic facts such that $F \vdash^R \Psi_d$, where Ψ_d is the derived fact represented by d.

3 Calculate State-Independent Activation Sets and Use Them in Relax-Plan

In our definition, an activation set of a derived fact consists of basic facts which can deduce the derived fact under the rule set R, regardless of any current state. Then, we present a depth-first algorithm (depicted in Fig. 3) to identify the set Σ which consists of all the state-independent activation sets (SIAS) in a given rule-graph.

```
Algorithm 1
SIAS-search (d, A, Path, Open)
Input: A derived fact (d), the state-independent
activation set under construction (A), the set of AND-
nodes of R on the search tree path from the search tree
root to d (Path), the set of nodes to visit for A (Open);
Output: The set of all state-independent activation sets
(Σ).

1. For each successor r of d do
2.     { Open ← Open ∪ {r};
3.       While Open ≠ ∅ do
4.         { x = first_element (Open);
5.           If x is a rule node Then Open ← Open ∪
   {first_successor (x) };
6.           Else If x is a basic fact Then
7.              { A ← A ∪ {x};
8.                If the antecedent of x has other
   unvisited successor x'
9.                Then Open ← Open ∪ {x'}
10.               Else Σ ← Σ ∪ {A};
11.              }
12.          Else If x is a derived fact Then
13.      If x ∉ Path Then
14.      { SIAS-search (x , A, Path ∪ {x} , Open);
15.          If the antecedent of x has other unvisited
   successor x'
```

Fig. 3. A depth-first algorithm to identify all possible SISA

```
16.              Then Open ← Open ∪ {x'};
17.      }
18.        }
19.For each x' in A which is the successor of r Do
20.        A = A - {x'};
21.}
```

Fig. 3. (*continued*)

The algorithm in Fig.3 performs a complete search on the rule graph. The function first_element gets the first element x in Open table. If x is a rule node, then one triggering condition of x enters the Open table. If x is a basic-fact node, then it becomes an element in A immediately. Until a triggering condition is totally supported by basic facts, the set A becomes an element in Σ. When x is a derived-fact node and doesn't emerge in the Path table, the search goes forward recursively by the line 14, otherwise, it should be pruned off to avoid cycle search. At last, to find all activation sets, the set A should maintain possible members (line 19~20). For example, we can get the set Σ_{SISA} of the derived fact "above(A,B)" on the rule graph in Fig.2 by this algorithm, as follow: {{on(A,B)}, {on(A,C), on(C,B)} , {on(A,C), on(C,D), on(D,B)} , {on(A,D), on(D,B)}, {on(A,D), on(D,C), on(C,B)}}. Here, {on(A,B)} is also a activation set in Σ_{SISA}, however, it belongs to the current state and hence doesn't appear in the set Σ_{LPG}. Next, we present a forward-search algorithm (depicted in Fig. 4) for the relaxed-plan in the domains with derived predicates, where the state contains not only basic facts, but also plenty of derived facts deduced from domain rules.

```
Algorithm 2
Extend-relax-plan (I, G)
Input: The initial state (I), the goal state (G);
Output: The set of actions or fail.
 1. S ← I ∪ D(I, R);
 2. level ← 0;
 3. For each action a which is applicable in S Do
 4.        S' ← S ∪ Add(a);
 5. level ← level + 1;
 6. S'' ← S' ∪ D(S', R);
 7. If S'' doesn't contain G , Then S ← S'' , GOTO 3
 8. Else π = extract-relax-plan(I, G);
 9. If π= Ø Then return fail;
10.Else return Aset(π).

Extract-relax-plan (I, G)
Input: The initial state (I), the goal state (G);
Output: The set of actions in relax-plan (Act).
```

Fig. 4. An algorithm to relax-plan with state-independent derived predicates

```
1. For each derived fact d in G Do
2.      G ← (G - {d})∪ best-SISA(d);
3. Act ← ∅;
4. select an action set A in the (level -1) layer in
   the relaxed plan graph, and A is a minimal set whose
   members can support furthest the basic facts of G ;
5. G ← (G- {add(a) | a ∈A}) ∪{pre(a) | a ∈A};
6. Act ← Act ∪ {A};
7. level ← level - 1;
8. If the level > 0 Then GOTO 1;
9. If G contains I Then return Act;
10.Else return ∅.
```

Fig. 4. (*continued*)

The algorithm in Fig.4 is composed of two phases: to extend the plan graph and to extract a plan solution. D(I, R) is the set of derived facts which is the closure of the set I under the rule set R (more details in [2]). Add(a) is the set of positive effects of an action node a, and pre(a) is the set of preconditions of an action node a. In the pre-process phase, we can first calculate the set Σ for every derived fact d by the function "SIAS-search (d, $\emptyset$, $\emptyset$, $\emptyset$)" and store it in a lookup table. And in relax-plan phase, each derived fact is replaced by its best state-independent activation set (SISA). By the way, a best SISA can be defined as an activation set which has the minimal actions set that can reach all basic facts from the initial state. By building a Lookup table to save state-independent activation sets, we can avoid spending a lot of time in calculating those state-dependent activation sets.

4   Experiments with FF-DP

These algorithms are implemented on a new planner, called FF-DP. We examined our planner in the benchmark domains: "PSR-middle" and "PROMELA-Optical Telegraph". And the experiment environment is CPU(Pentium Processor 1.2G) + RAM (512M) + Redhat9.0 Linux, with the compiler gcc 3.0. Actually, we find that the performance of FF-DP is satisfactory in most specific problems. In Fig.5[1], when the problem is relatively simple, FF-DP performs a little worse than LPG-td, for the time spent in preprocess phase by the former is a little more than the recalculation spent by the latter. However, when the problems become more and more complex (that is, contain more derived rules), such as pfile39, 41, 45 or 49, the time of plan by FF-DP is much shorter than that by LPG-td. Also, we are confused that why SGPlan performs so erratically. On the other hand, in Fig.6, we can see that the performance of LPG-td is very bad because of some unexpected errors in the program when in IPC-4 events, while the performance of FF-DP is competitive with the Downward planner, which introduces artificial actions and facts to compile derived predicates away. From these results, we can see that our idea of state-independent activation sets and the corresponding algorithms are effective and promising.

---

[1] On the x-axis we have the problem name (abbreviated by numbers). On the y-axis, we have CPU-time (log-scale).

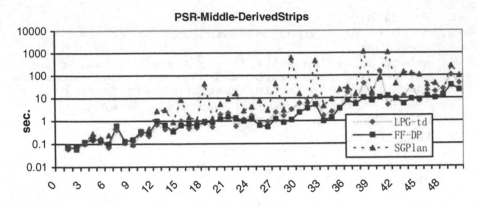

Fig. 5. Performance of FF-DP in the domain PSR-Middle

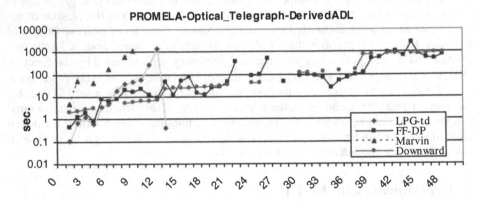

Fig. 6. Performance of FF-DP in the domain PROMELA-Optical Telegraph

5 Conclusion

Handling derived predicates efficiently in modern automated planners is an important problem with a lot of recent practical interest in the AI planning community. Derived predicates can't appear in any effect of action and their truth value can only be deduced from domain rules. In this paper, we propose state-independent activation sets to avoid a number of calculations when the current state continuously changes. Two algorithms are presented: one for calculating the set of state-independent activation sets of a derived predicate in a given rule graph; and the other is to use these activation sets to replace derived predicates in a relaxed plan. We implement these algorithms on a new planner, called "FF-DP", and find good results in the domains containing derived predicates. Future work includes further defining the best activation set and discussing how it will affect on selecting solution sequences.

References

[1] S. Edelkamp, and J. Hoffmann. PDDL2.2: The language for the Classic Part of the 4th International Planning Competition. T.R. no. 195, Institut f'ur Informatik, Freiburg, Germany, 2004.

[2] S. Thiebaux, J. Hoffmann, B. Nebel. In Defense of PDDL Axioms. Artificial Intelligence. Volume 168 (1-2), 2005, pp. 38 - 69.

[3] J. Hoffmann, S. Edelkamp, R. Englert, F. Liporace, S. Thiebaux, and S. Trueg. Towards Realistic Benchmarks for Planning: the Domains Used in the Classical Part of IPC-4 - Extended Abstract. Proceedings of the 4th International Planning Competition (IPC-40). 2004, June, Whistler, Canada, pp. 7-14.

[4] M. Davidson, M. Garagnani. Pre-processing planning domains containing Language Axioms. In Grant, T., & Witteveen, C. (eds.), Proc. of the 21st Workshop of the UK Planning and Scheduling SIG (PlanSIG-02). pp. 23-34. Delft (NL), Nov. '02.

[5] Alfonso Gerevini, Alessandro Saetti, Ivan Serina, Paolo Toninelli. Planning with Derived Predicates through Rule-Action Graphs and Relaxed-Plan Heuristics. R.T. 2005-01-40.

[6] Jörg Hoffmann, Bernhard Nebel. The FF Planning System: Fast Plan Generation through Heuristic Search. Journal of Artificial Intelligence Research 14 (2001), pp.253-302.

Elicitation of Non-functional Requirement Preference for Actors of Usecase from Domain Model

G.S. Anandha Mala[1] and G.V. Uma[2]

Department of Computer Science and Engineering, College of Engineering,
Anna University, Guindy, Chennai, Tamil Nadu, India-600025
malamanosuke@yahoo.co.in, gvuma@annauniv.edu

Abstract. Requirement engineering plays a vital role in the development of the software. The quality of the software being developed depends on the non-functional requirements, which are still not derived effectively due to the conflicts between them. This paper presents an approach to identify the non-functional requirements for a given usecase description from the domain model such as Unified Modelling Language class diagram and goal based questionnaires. This approach makes use of the domain model to find out the behaviour of the system and possible constraints for actors in the system. The non-functional requirement taxonomy and the user preferences are used to analyse the conflicts, which is resolved based on trade-off analysis by prioritizing the preference. The prioritization depends on the dominating non-functional requirements from the inference engine.

1 Introduction

The objective of this system is to provide a support to requirements engineering to identify non-functional requirements of the actors with their preference. This procedure analyses the events triggered from the use case description to find out the variants in the requirements. This is performed using usecase diagram, domain model and taxonomy of non-functional requirements along with actor's preferences. The proposed methodology is explained in section 2. Our implementation and results are explained in section 3. The conclusion and future work is contained in Section 4.

2 The Proposed System

Haruhiko Kaiya [8] discusses the elicitation of the non-functional requirement performed by comparing the existing usecase to derive the invariants related to the non-functional requirements and the stakeholders involved. The system does not focus on the internal description of the usecase as well as the quality requirements are not prioritized. The system proposed in this paper considers the usecase description to identify the system interaction by comparing it with the domain model; also trade-off analysis is used to prioritize the quality requirements. In the earlier works [3][4][7][9][10], domain modeling and extracting information from usecase models

A. Hoffmann et al. (Eds.): PKAW 2006, LNAI 4303, pp. 238–243, 2006.

have done separately. We have suggested a method of combining both, and extracting the variants in usecase and combining with domain model and non-functional taxonomy to derive the actor's preference. The system architecture is shown in Fig. 1.

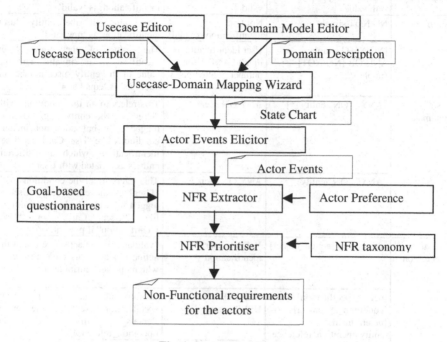

Fig. 1. System Architecture

Usecase and Domain model are structured using the XML editor to the Data Type Definition (DTD) format. The editor checks the syntax of both the usecase and domain model with the specified structure. They are represented with specified notations for easy traceability of the usecase description with the domain model. The syntactic structure of the usecase description is validated using natural language processing. The various syntactic structures used in usecase and domain model are listed in the Table 1.

The Usecase-Domain Mapping Wizard extracts the entities, which are not present in the domain model from the usecase description by mapping the usecase with domain model. The usecase follows the below structure.

Title: a label that uniquely identifies the use case within the usecase model.
Primary Actor: the actor that initiates the use case.
Participants: other actors participating in the use case.
Goal: primary actor expectation at the successful completion of the use case.
Precondition: condition that must hold before an instance of usecase can be executed.
Postcondition: condition that must be true at the end of a 'successful' execution of an instance of the use case.
Steps: Sequence of steps involved in the usecase along with extension.
Extensions: a set of step extensions that applies to all the steps in the use case.

Table 1. Syntactic Structure Used in Usecase and Domain model

Statements	Syntax	Sample	Description
simple	[Determinant] entity[1] verb value	User identification is valid	The value of the entity 'user identification' is 'valid'
complex	NO/NOT simple	Not User identification is valid	The value of the entity 'user identification' is 'not valid'
	[NO/NOT] simple AND/OR [NO/NOT] simple	User identification is invalid AND User number of attempts is equal to 4	The value of the entity 'user identification' is 'invalid' and the value of the entity 'user number of attempts' is 'equal to 4'
ANY statement	"ANY" "ON" entity [*]	ANY ON user*	This refers to all the conditions with "User" as the entity (e.g. "User is logged in"), but does not include conditions like "User Card" or "User identification" which are different entities associated with User
	"ANY" "ON" entity	ANY ON user	This refers to all the conditions with "User" as the entity (e.g. "User is logged in"), also includes conditions like "User Card" or "User identification" if present.
Operation declaration	action_verb [action_object]	Validate user identification	"validate" is the action verb and the action object is "user identification" which is an attribute of concept "User".
	[delay_specification] [condition_statement] [determinant] entity operation_reference	BEFORE 60 sec, USER enters pin	'Before 60 sec' is the delay specification,'User' is the entity ,'enters pin' is the operation_reference
		ATM asks user validation to the Bank	ATM is an entity. 'asks user validation to the bank' is operation_ reference
operation_reference	conjugated_action_verb[2] [(binding_word[3])+] action_object [action_participant]	asks user validation to the bank	'Validation' is the conjugated action verb, 'to' is the binding word, 'bank' is the action_object
condition_statement	"IF"simple/complex "THEN"	IF User Identification is valid THEN, ATM displays operation menu	The simple statement 'User identification is valid' is taken as the condition
branch	[delay_specification] [condition_statement] "GOTO"["STEP"] step_reference	Go to Step 2	Control transferred to step number 2

[1] An *entity* consists of one or more words specified as $Word_1, \ldots word_n$. The sequence of words must correspond to a concept, an attribute of a concept in the domain model, an instance of a concept, or a reference to an attribute of an instance.

[2] conjugated_action_verb is the action_verb used in the concept operation declaration in the present tense.

[3] binding_word may be a possessive adjective, article or preposition.

Also it captures the non-existing actors, operations and conditional statements. Then updates the domain model using the reverse engineering wizard supported by the UML plug-in. The wizard uses the structured usecase and domain concept. Usecase editor and domain editor are used for this purpose.

'Actor Event Elicitor' maps the Usecase with the domain model based on the pre-conditions of the usecase. On successful completion the precondition states are withdrawn. The state chart for the entities, which are mapped with the domain model, is generated then. The state chart contains entries like "*1 ---[insert card]--> 2*", meaning that from state 1 it goes to state 2 on performing the event 'insert card'. From the state chart the events related to the actors alone are identified.

'NFR Extractor' generates the non-functional requirements for the actor events with the help of the actor preference and goal-based questionnaires. Actor preference is a matrix of actor versus event. The matrix entries tell whether the actor can perform the specified event or not. In goal-based questionnaires all the events are embedded with possible questions, which helps to identify the quality requirements. Sample actor preference matrix and goal-based questionnaires are shown in the Table 2 and Table 3 respectively.

Table 2. Actor Preference Matrix

Events	Patient	User	Doctor	Nurse
Triggers alarm	Yes	No	Yes	Yes
Press logout button	Yes	Yes	Yes	Yes
Enter vital signs	Yes	No	Yes	Yes
Insert Card	Yes	Yes	Yes	Yes
Connect cables	No	No	Yes	Yes

Table 3. Goal-based questionnaires

Events	Preference	NFR
Insert card	Card Expired	Security
Enter pin	Invalid pin number	Security
Enter pin	Provide proper human computer interface	Usability

'NFR Prioritizer' prioritizes the identified non-functional requirements based on the trade-off analysis. The results are shown in Fig. 2. This is done by the inference engine, which in turn makes use of the NFR taxonomy. In NFR taxonomy, all the NFRs are associated with other conflicting and dependable NFRs. The NFR taxonomy looks like,

Usability#Simplicity+#Accessibility+#Installability+#Operability+#Maintainability-

It states that simplicity, accessibility, installability and operability are directly proportional and maintainability is indirectly proportional with usability.

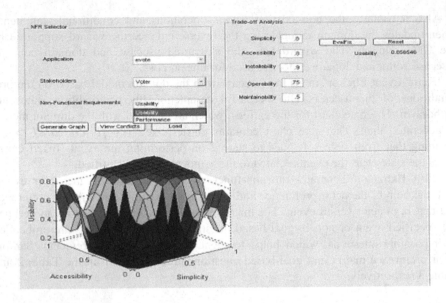

Fig. 2. Trade-Off Analysis

3 Implementation and Results

The system has been implemented using JAVA and a plug in to eclipse for reverse modelling. UML interface 'nsuml' is used to generate the state machines. This is used to synthesize the user behaviour of the system. The editor is created for generating both use case description and domain model description. The system is tested for the following domains ATM, Retailing system, Patient Monitoring System and E-voting system. The system will act according to the user's preference given in the non-functional requirement taxonomy.

The usecase and domain model makes use of XML editor to create data type definition (DTD) to store the model. The XML reader and writer are used to import and export the files for processing. The use case and domain model make use of structured text, which has been checked for the syntax. The wordnet is used as an interface to check for valid parts of speech.

Domain model makes use of UML diagram descriptions. The reverse engineering wizard is used to check usecase and domain maps. Workbench.jar, jface.jar, jdom.jar, runtime.jar are all used to implement the domain mapping and extracting. The prioritization of the non-functional requirement is mapped to goal questions, which are stored in Microsoft Access database. The conflicts are stored in a separate data file. The conflicts are resolved using inference engine created in matlab. The inference engine calculates the preference from the given user weights for each non-functional requirement.

The inference engine is designed to perform trade-off analysis for the non-functional requirements such as usability, performance, maintainability, security, correctness, authorization, reliability and availability.

4 Conclusions and Future Work

This paper presents an approach for deriving non-functional requirements by comparing usecase description with the domain model. This system takes a usecase written in a restricted form of natural language and generates a state model that integrates the behaviour specified by the usecase. The invariants and its initiators are captured from the usecase and domain model. The conflicting non-functional requirements are derived from the NFR taxonomy and they are prioritized using trade-off analysis. The system can be extended to automate the trade-off analysis by using an intelligent system.

References

1. Markus Nick, Klaus-Dieter Althoff, Carsten Tautz: Facilitating the Practical Evaluation of Organizational Memories Using the Goal-Question-Metric Technique. KAW'99 – Twelfth Workshop on Knowledge Acquisition, Modeling and Management 1999.
2. Jane Cleland-Huang, Raffaella Settimi, Oussama BenKhadra, Eugenia Berezhanskaya, Selvia Christina: Goal-Centric Traceability for Managing Non-Functional Requirements. ACM ICSE'05 May 15–21, 2005
3. Annie I. Anton, Colin Potts: The Use of Goals to Surface Requirements for Evolving Systems. 20th International Conference on Software Engineering (ICSE98), pages 157-166, April. 1998
4. Luiz Marcio Cysneiros, and Julio Cesar Sampaio do Prado Leite: Nonfunctional Requirements: From Elicitation to Conceptual Models. IEEE Transactions On Software Engineering, Vol. 30, No. 5, May 2004
5. Xiaoqing Frank Liu, John Yen: An Analytic Framework for Specifying and Analyzing Imprecise Requirements. Proceedings of 18th International Conference on Software Engineering (ICSE-18), Berlin, Germany, pp 60-69, March 25-30, 1996
6. Martin Glinz: Rethinking the Notion of Non-Functional Requirements. Proceedings of the Third World Congress for Software Quality (3WCSQ 2005), Munich, Germany, Vol. II, 55-64
7. Vittorio Cortellessa, Katerina Goseva-Popstojanova, Ajith R. Guedem, Ahmed Hassan, Rania Elnaggar, Walid Abdelmoez, Hany H. Ammar: Model-Based Performance Risk Analysis. IEEE Transactions On Software Engineering, Vol.31, No.1, January 2005
8. Haruhiko Kaiya, Akira Osada, Kenji Kaijiri: Identifying Stakeholders and Their Preferences about NFR by Comparing Use Case Diagrams of Several Existing Systems. Proceedings of the 12th IEEE International Requirements Engineering Conference (RE'04) IEEE 2004
9. John Yen, W. Amos Tiao, and Jianwen Yin: STAR: A Tool for Analyzing Imprecise Requirements. *Proceedings of 1998 IEEE International Conference on Fuzzy Systems (FUZZ-IEEE '98)*, Anchorage, Alaska, pp. 863-868, May 4-9, 1998
10. Andreas Gregoriades and Alistair Sutcliffe: Scenario-Based Assessment of Nonfunctional Requirements. IEEE Transactions On Software Engineering, Vol. 31, No. 5, May 2005

Enhancing Information Retrieval Using Problem Specific Knowledge

Nobuyuki Morioka and Ashesh Mahidadia

School of Computer Science and Engineering,
The University of New South Wales,
Sydney, Australia
{nmor250, ashesh}@cse.unsw.edu.au

Abstract. In the recent past, information retrieval techniques have improved significantly and it is now possible to access massive text corpora using some of the most popular search engine tools like Google, Yahoo, PubMed[1] (popular search engine for medical literature), etc. Considering that such search engine tools are normally trying to retrieve information from massive text corpora, a number of search results they need to display might be in hundreds or even in tens of thousands. Normally it is not possible or practical to browse through a very large collection of search results, and often a typical user needs further assistance in focusing on the search results that might best meet his/her requirements.

In this paper we present a new technique that allow a user to interactively express problem (task) specific knowledge (which is otherwise not possible using search engine tools like Google, Yahoo, PubMed, etc) and later use this knowledge to help a user to interactively and quickly focus on search results they might be interested in. The system presented in this paper integrates some of the techniques from the field of Natural Language Processing and Visualisation.

Keywords: Intelligent Systems, Information Retrieval, Knowledge based Information Retrieval.

1 Introduction

In this digital age it is now possible to access massive amounts of information available online from almost anywhere in the world. The challenge now is to get the most relevant information quickly and easily. To address this challenge, in the recent past information retrieval techniques have improved significantly and we now have access to some of the most popular search engines like Google. Yahoo, PubMed (popular search engine for medical literature), etc. However, considering that such search engines are normally trying to find relevant information from a very large database(s), often they return a large number of relevant documents for a given query. Normally these documents (returned hits) are ranked based on their relevance to the

---

[1] http://www.pubmed.gov

A. Hoffmann et al. (Eds.): PKAW 2006, LNAI 4303, pp. 244–251, 2006.

query and displayed in non-increasing order. Search engines typically return thousands or even hundreds of thousands of relevant documents for a given query, and it is then up to a user to browse through a large collection of relevant documents and look for documents that meet her/his criteria. Users normally browse through the first few pages of the high-ranking documents (say 2 or 3 pages of search results, each page with 10 search results) and if they do not find what they are looking for, they quickly start losing interesting in the search results. We believe here a user needs some assistance to quickly and easily focus on the search results that might meet his/her requirements.

In this paper we present a new technique that acquires problem (task) specific knowledge from a user, and later uses this knowledge to help the user to interactively and quickly focus on search results that might be of interest. The rest of the paper is organized as follows: Section-2 briefly outlines some of the relevant features of the current Information Retrieval (IR) systems. Section-3 describes our approach and discusses how it could enhance Information Retrieval. Section-4 describes experimental results and discusses how our approach could enable a user to quickly focus on search results that are of interest. In Section-5 we present our discussion and conclusions.

2 Information Retrieval Systems

Information retrieval activity typically starts with an *information need* of a user. A user then expresses his/her information need in terms of a *query* in order to find relevant documents. The query expressed by the user may not be the best articulation of the information need, however it is usually the only clue that the search engine has concerning the user's goal [2]. Recently, search engines have tried to address some of these problems by using a technique known as query expansion [3]. In this technique, similar or related terms are also considered while looking for relevant documents. For example, a search engine could find synonyms using a thesaurus or related terms using a domain dependent ontology [10] or from other relevant documents (also known as relevance feedback) [4]. Once the query terms are finalized, a search engine tries to find relevant documents based on the query terms and usually displays results in non-increasing order of relevance.

In the above traditional view of IR, a user is limited to expressing his/her information need in terms of keywords that best describe the requirements. However, it is possible that for a specific retrieval task (problem), the user expects the required keywords to be at or near the beginning of the document, or may be at the middle or at the end of the document. This would be particularly true if a user roughly knows the structure of a document(s) he/she is interested in. For example, a user might be searching for a tutorial on prolog. Based on the past experience, the user knows that a typical page would have the words "prolog" and "tutorial" at or near the beginning of the document. However, the user might not be sure about the sequence in which they might appear. For example, it could be "Prolog Tutorial" or "Tutorial on Prolog" or more importantly it might be divided into more than one lines like: two lines "Week-12 Tutorial" and "Prolog" separated by say 1 or 2 lines, at or near the beginning of the document. Unfortunately the current IR systems are not able to effectively use such problem (task) specific knowledge. They do offer features like searching keywords in the Title, URL, etc. However, these features do not cover the types of cases we outlined above.

In general, it is fair to say that the current IR systems are primarily focused on improving retrieval algorithms, and they pay less attention to the task of properly acquiring information need of a user in the first place [1, 5, 6, 7]. We believe that future IR systems should be knowledge-driven where an information retrieval process is enhanced by properly acquiring information need and problem specific knowledge from a user.

Most of the current IR systems use very similar (and one could even say traditional) form based user interfaces [1]. Considering that many users today have access to powerful machines, we believe more advanced interactive user interfaces should be explored in order to improve IR experiences. However, there are very few systems [8, 9] today that try to use such advance user interfaces in order to improve IR experiences.

3   Using Problem Specific Knowledge to Enhance Information Retrieval

In this section we describe the system that addresses some of the problems discussed in the previous section. Fig 1 shows the overall system architecture.

Search Engine Module: A user starts using the system by initially providing a (traditional) query and a search engine name to the system. The "Search Engine" module simply executes the query on the nominated search engine; fetches search results; and later fetches the corresponding documents from the Web. It is obvious that a user can use all the available advance features of a nominated search engine here.

Problem Specific Knowledge (PSK) Module: In addition to a (traditional) query, a user can also ask the system to generate problem specific attributes from the documents obtained by the module "Search Engine". For example, a user can ask the system to find the first occurrence of the word "tutorial" (say ignoring case) in every document retuned by the module "Search Engine". Similarly, we can also ask the system to find the first occurrence of the word "prolog" (again say ignoring case), and so on. The module named "Problem Specific Knowledge (PSK)" allows users to interactively specify such problem specific attributes. Currently this module support the following positional attributes:

- First occurrence of a word (in token, sentence and/or paragraph number)
- Last occurrence of a word (in token, sentence and/or paragraph number)
- First occurrence of a group of words (in sentence and/or paragraph number)
- Frequency of a word

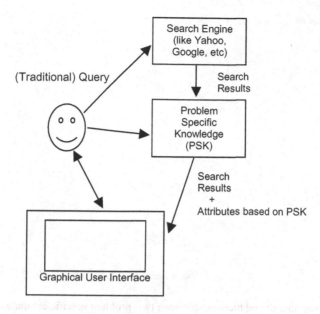

(Traditional) Query

Search Engine
(like Yahoo,
Google, etc)

Search
Results

Problem
Specific
Knowledge
(PSK)

Search
Results
+
Attributes based on PSK

Graphical User Interface

Fig. 1. Overall System Architecture

In the near future, we plan to extend this list by including more commonly used attributes from the field of Natural Language Processing (NLP) [2]. However, as discussed in the next section, we believe the above attributes are powerful enough to significantly enhance IR experiences for a large number of problems.

The module PSK uses GATE[2] to generate (calculate) NLP related attributes. GATE (General Architecture for Text Engineering) is a widely used and very popular (free) open source framework (or SDK) that is successfully used for all sorts of language processing tasks.

Graphical User Interface Module: The problem specific attributes generated by the module PSK, along with the corresponding documents are sent to the "Graphical User Interface" module. The aim here is to plot charts based on the values of the problem specific attributes. For example, we can plot a chart where an X-axis represents first occurrences of the word "tutorial" and Y-axis represents first occurrences of the word "prolog". By plotting such charts, a user can locate documents where these values are either say very small or very large (depending on the requirements). Alternatively, a user may not be interested in documents where one value is say small and another say high, and so on. The module allows a user to quickly plot such charts to view documents in a variety of ways. Each point on a chart represents a document, and it has a hyperlink to the corresponding document. In other words, a user can simply click on a point in a chart and the corresponding document will be displayed in a browser window. This would allow a user to interactively browse through a large collection of search results, using problem specific attribute values.

---

[2] http://gate.ac.uk/

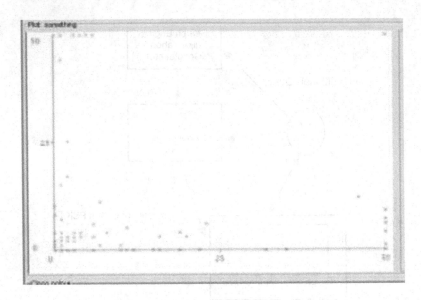

Fig. 2. Chart outlining the relationship between two problem specific attributes: prolog-1p (X-axis) and tutorial-1p (Y-axis)

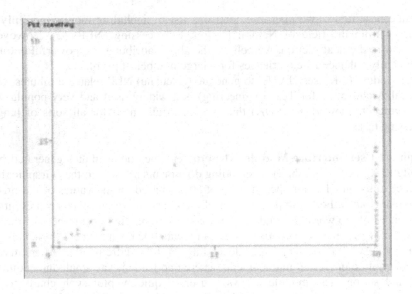

Fig. 3. Chart outlining the relationship between two problem specific attributes: prolog-tutorial-1p (X-axis) and prolog-1p (Y-axis)

It should be noted that here the whole process is knowledge-driven. A user defines problem specific attributes and he/she also selects attributes for charting a graph. This is different to classical clustering methods (which are data-driven) where documents are grouped together using predefined clustering criteria. In this paper we will only focus on knowledge-driven explorations. Here we simply note that if required, it is also possible to cluster documents based on their problem specific attributes for data-driven explorations. The module uses Weka's Visualisation tool[3] to display charts. Considering that Weka offers tools for popular data mining techniques, it would not be difficult to also include data-driven approaches in the future. However, in this paper we want to emphasis knowledge-driven approaches that are often neglected in IR literature.

4 Experimental Result

In this section we present experimental results that demonstrate how problem specific attributes could be used for knowledge-driven explorations of retrieved documents. For the following query, we use Yahoo search engine, retrieve the first 100 results, and restrict our search to edu.au (to avoid possible advertisement material).

Let's first continue our previous example where we were interested in searching for a prolog tutorial. For this task, we created the following attributes:

- Paragraph number where the word "prolog" (ignoring case) appears first time in a document, let's call it **prolog-1p**
- Paragraph number where the word "tutorial" (ignoring case) appears first time in a document, let's call it **tutorial-1p**
- Paragraph number where both the words "prolog" (ignoring case) and "tutorial" appear first time in the same paragraph, in a document, let's call it **prolog-tutorial-1p**

The form-based graphical user interface allows a user to create the above attributes. After calculating the above attribute values, the system sends these attribute values along with document references to Weka's Visualisation tool (which is appropriately modified for the system). Initially the system displays thumb nails for possible charts. Here a user can quickly examine different thumbnails to look for possible interesting patterns (relationships) between problem specific attributes. By clicking on a thumbnail, a user can display the corresponding chart. Fig 3 shows such a chart for the attributes prolog-1p and tutorial-1p. Similarly, Fig 4 shows relationships between attributes prolog-tutorial-1p and prolog-1p.

Given that we are looking for a prolog tutorial, we might be more interested in exploring documents that are close to the origins (bottom-left corner) of these charts. We could also infer other useful information from theses charts. For example, in Fig 3, documents that appear on the diagonal axis or close to the diagonal axis have both the attributes appearing first in the same paragraph or nearby paragraphs, increasing the likelihood of them being prolog tutorial. Alternatively we could say that in Fig 3 documents with low x-value (prolog-1p) and high y-value (tutorial-1p) might be referring to some other material on prolog (like lecture notes on AI). Similarly, we

---

[3] Weka is an open source software for data mining and visualisation, available at http://www.cs.waikato.ac.nz/ml/weka/

could say that documents with high x-value (prolog-1p) and low y-value (tutorial-1p) might be referring to tutorials on other topics (again say AI tutorials).

In the charts there are some documents with attribute values 50. In this experiment, if we cannot find the required term in the first 10,000 words (or there is a parsing problem), we assign 50 to that attribute. This is to indicate that the corresponding value is too high for our purposes. We did this to simplify chart displays. Also note that, a paragraph index starts with 0 and a Yahoo rank starts with 1.

We manually checked all the 100 documents returned by Yahoo! for their relevance to our task and marked them as a relevant or not. Out of the 100 documents retrieved, there are 35 (35%) relevant documents for the task. Based on the Yahoo Ranking, 70% of the top 10 documents are relevant, 65% of the top 20 documents are relevant and 53.3% of the top 30 documents are relevant.

In Fig 2, there are 26 documents on or near the diagonal axis. Out of these 26 documents where an absolute difference between the paragraph indexes is less than 1, there are 16 relevant documents, that is 61.5% relevant documents. 11 of these 16 relevant documents have Yahoo ranks greater than 30. In other words, these 11 relevant documents would not appear in the first three pages of the search results. This is useful because a user can browse the top ranking Yahoo hits (say first few pages) and then use charts to look for more relevant documents, or vice versa.

In Fig 3, there are 20 documents that appear near the origin. In other words, these are the documents where both the words appear in one paragraph at or near the beginning of the documents. Out of the 20 such documents, 65% of the documents are relevant. Out of these 20 relevant documents, there are 5 documents with Yahoo ranks greater than 30.

5 Discussion and Conclusions

Popular search engines like Yahoo do use word proximity as one of their retrieval criteria. However, the final ranking is based on multiple criteria and hence it is not easy to identify documents with a specific structure or that satisfy a specific criterion. The approach presented in this paper enhances a retrieval process by combining problem specific knowledge with the underlying strength of a given search engine. In the experiments presented above, the criterion used in Fig 2 is more relaxed than Fig 3, and therefore we believe there are more relevant hits for Fig 2. This might also be the reason why there are more relevant documents in Fig 2 with Yahoo ranks greater than 30.

In summary, we believe that the current IR systems do not consider a role of a user very seriously in defining information need, and later navigating through possible solutions. More research is need to design and develop innovative approaches that keep users in the loop and actively seek more domain (problem) specific knowledge that could be effectively used in narrowing the search space or improving matching criteria. The system presented in this paper is just the beginning of a bigger goal of building a smart interactive information retrieval system.

References

1. Belkin, N., et al; Evaluating Interactive Information Retrieval Systems: Opportunities and Challeges. in Conference on Human Factors in Computing Systems (CHI'04). 2004: ACM Press, New York, USA.
2. Jackson, P. and I. Moulinier; Natural Languare Processing for Online Applications. 2002: John Benjamins Publishing Company.
3. Ruthven, I.; Re-examining the Potential Effectiveness of Interactive Query Expansion; in SIGIR 2003: Proceedings of the 26th Annual International ACM SIGIR Conference on Research and Development in Information Retrieval. 2003. Toronto, Canada: ACM.
4. Salton, G. and C. Buckley; Improving retrieval performance by relevance feedback; Journal of the Americal Society for Information Science, 1990. 41(4): p. 288-297.
5. Shen, X., B. Tan, and C. Zhai; Context-Sensitive Information Retrieval Using Implicit Feedback; in Proceedings of the 28th annual international ACM SIGIR conference on Research and development in information retrieval SIGIR '05. 2005. Brazil: ACM Press.
6. Stojanovic, N.; On the Role of a User's Knowledge Gap in an Information Retrieval Process. in Proceedings of the 3rd International Conference on Knowledge Capture (K-CAP 2005). 2005. Banff, Alberta, Canada: ACM.
7. Voorhees, E.H. and D. Harman; Overview of the sixth text retrieval conference (TREC-6). in Information Processing and Management,. 2000.
8. KartOO visual meta search engine [http://www.kartoo.com]
9. Vivisimo's clustering [http://www.vivisimo.com]
10. Müller HM, Kenny EE, Sternberg PW (2004); Textpresso: An Ontology-Based Information Retrieval and Extraction System for Biological Literature. PLoS Biol 2(11): e309

Acquiring Innovation Knowledge

Peter Busch and Debbie Richards

Computing Department,
Division of Information and Communication Sciences,
Macquarie University, Australia
{busch, richards}@ics.mq.edu.au

Abstract. There are few possibilities for acquiring knowledge related to innovation. Firstly, acquiring knowledge using machine learning typically requires structured and classified data and/or cases, and lots of them. Secondly, manual acquisition of knowledge requires human expertise. Both approaches seem impractical when it comes to innovation knowledge. While innovation is recognized as a vital part of sustainability within organizations, there is little assistance with how we can acquire, reuse or share the innovation knowledge that may exist. We suggest a technique and present preliminary results of an evaluation study using this approach.

Keywords: Innovation knowledge, knowledge acquisition.

1 Introduction

Many today would accept that the Western organisation is no longer competitive from the point of view of secondary industry. Although both primary and tertiary industry must be conducted onshore, to attain a global advantage at the quaternary and quinary levels requires innovation. Naturally attaining a competitive advantage is easier said than done for "innovation is... a significant and complex dimension of learning in work, involving a mix of rational, intuitive, emotional and social processes embedded in activities of a particular community of practice" [5, p.123]. We too see innovation taking place as a process whereby knowledge may be gained either through self experience over time or by serving in an 'apprenticeship' with a more experienced innovator who may pass some of his or her expertise on. Nevertheless innovation is not simply a process of trial-and-error rooted in experience, innovation needs to produce timely and ongoing results "involving a complex mix of tacit knowledge, implicit learning processes and intuition" [5, p.124]. Given the acknowledgement of the connection between tacit knowledge and innovation knowledge [9], we have turned our research using work-place scenarios to capture tacit knowledge toward the capture of related innovation knowledge.

2 The Approach

The approach carries on and extends our previous work [2, 3, 4] with a narrowing of focus to innovative and creative type knowledge and a change of direction into the

A. Hoffmann et al. (Eds.): PKAW 2006, LNAI 4303, pp. 252–257, 2006.

application of personnel recruitment and training. Acknowledging that innovation is a process we will be looking for emerging patterns of behaviour appropriate to each of the various phases of the innovation process and how these responses correspond to our current understanding of innovation including the various psychological models, instruments and approaches which exist.

Similar to our previous work in developing an IT Tacit Knowledge inventory along the lines of Sternberg *et al.* [11], we have established an inventory with twelve randomly assigned 'innovation' scenarios. We see an example of scenario 12 with corresponding answer 'options' in Fig. 1. For each of these answers respondents select **two** Likert scale values (Extremely Bad through to Extremely Good) for **both** how they would *ideally/ethically* rate the answer option, **and** *realistically* how they feel the answer option is with regard to dealing with the given scenario. We also want our respondents to add innovative scenarios and answer options of their own with a view toward extending the inventory for future use. Finally, we ask our sample population to select the stage of innovation of the scenario along the lines of the Novelty Generation Model (NGM) [10].

You work for an internet company whose founder and chief executive routinely abuses and demoralises people. You and your fellow employees dread coming to work with this tyrannical executive, but you know that he has a great idea that can be packaged for a hot initial public offering in the next 12 months.

Do you:

1) Wait until the company goes public and its stock options vest then get out of there as quickly as possible.

2) Reduce annual leave and join another company. You don't have to take that kind of abuse.

3) Steal his idea and make some subtle readjustments to make it better then start your own internet company. With any luck you'll be able to bankrupt him and make a lot of money in the process.

4) Stay with the company for as long as they'll have you. Company loyalty is always appreciated, and the executive's ideas have merit even if he is a jerk.

5) Approach the chief executive and tell him firmly but politely that you don't appreciate his behaviour towards you and the rest of his staff.

6) Don't take his insults lying down, rise to the occasion and return them with interest.

7) Try to find out what the executive's real problem is. It may turn out to have nothing to do with you and rather be connected to personal problems. In which case, you won't feel that you are incompetent at your job.

Fig. 1. Scenario 12 with associated answers

The NGM is a bio-psycho-social approach, for it recognises that at a genetic level some people are more inclined to look for new problems and able to come up with novel solutions. In the model, the first step is novelty seeking followed by creativity which is broken into novelty-finding and novelty-production. These stages may be divided along the following lines.

Idea generation: Typically a technical insight into a product or process or thought about a service.

Opportunity recognition: An opportunity is identified for developing an idea into a new product, process or service.

Development: Usually involving prototype development and marketing testing.

Realisation: Typically realising how to market the product and introduce it to the customer.

These stages relate to novelty seeking with idea generation being a form of novelty seeking, opportunity recognition comprising novelty finding, and the last two stages representing a form of production from a novel idea. We return to these stages shortly.

Whilst we recognize certain psychological approaches such as the Kirton Adaptation-Innovation (KAI) inventory [8] or the Myers-Briggs Type Indicator (MBTI) creativity index [7] also focus on innovators, we choose to focus more so on the behaviour of individuals who have had successful results rather than on character or personality traits that so typically characterise current psychological research. However, we envisage that such psychometric tests will also play a role in a comprehensive instrument that can be used for the recruitment and development of personnel.

3 An Evaluation Study

As Information and Communication Technology (ICT) is our area of expertise, we will initially focus on innovation in this field. To compare novices with expert innovators, we are using two sample populations. First of all a third year undergraduate 'management theory' class of 75 individuals with a median age of 21 forms our novice population, and secondly approximately a dozen recognized innovators varying from 30 to 80 years of age, who will provide a skilled sample data set to compare against. To be recognised as an innovator, as opposed to merely claiming to be one, infers a process of public scrutiny. The individuals we will be approaching will by definition generally fit within the category of people experienced at what they do. With the incorporation of biographical information into the first component of the inventory, we hope to find differences in the answering of the scenarios on the basis of gender, or employment seniority, LOTE (Language Other Than English), highest formal qualification obtained and amount of ICT experience. Naturally the last two factors will not be high for the novice group given the age group we are dealing with.

4 Results and Findings

We present only a very small selection of our results here to illustrate our technique. Our novice population is 20 to 26 years of age, largely male (only 5 females), overwhelmingly ethnic (where ethnic in the Australian context refers to non Anglo-Celtic) and more specifically concentrated in the Chinese and to a lesser extent, the sub-continental ethnic groups. Finally the novices were generally school leavers (highest qualification was typically completion of secondary school) as one would expect. Analysis of the results revealed that all respondents took the innovation knowledge inventory seriously and none took a neutral 'Neither Good nor Bad' Likert scale option all the way through the questionnaire. To maintain concentration and thereby increase data validity, respondents were given only 4 randomly assigned scenarios along with the biographical component of the questionnaire.

Let us briefly examine the results of the answers for part of the inventory, in this case for scenario 12. With regard to answer 1 ("Wait until the company goes public and its stock options vest then get out of there as quickly as possible"), our respondents were ethically generally ambivalent, hovering around neither good nor bad, but realistically this option was considered on the whole to be good idea.

With regard to answer 2 ("Reduce annual leave and join another company. You don't have to take that kind of abuse"), the respondents were ethically positive, but realistically more negative. In other words whilst this option might seem an okay thing to do, our respondents felt in practice this was not such a good idea.

Answer 3 for Scenario 12 ("Steal his idea and make some subtle readjustments to make it better then start your own internet company. With any luck you'll be able to bankrupt him and make a lot of money in the process") presents the most interesting result. There is clearly a *very* strong skew toward answering this question in the negative from an ideal or ethical point of view, but our undergraduates feel in practice this option is not so bad with a small majority actually considering the idea positive in practice.

With regard to answer 4 ("Stay with the company for as long as they'll have you. Company loyalty is always appreciated, and the executive's ideas have merit even if he is a jerk"), our novice population is evenly spread with regard to this situation from an idealistic point of view. In practice however the novices are inclined toward considering this option a bad idea.

In answering 5 ("Approach the chief executive and tell him firmly but politely that you don't appreciate his behaviour towards you and the rest of his staff"), the undergraduates feel this is a very good idea idyllically speaking. In practice however, they seem a little more reserved, a small minority even considering this an extremely bad idea.

Answer 6 ("Don't take his insults lying down, rise to the occasion and return them with interest") is taken on the whole negatively by our sample students. What is interesting is that a larger than usual group of 'fence sitters' take a neutral stance ('Neither Good nor Bad') for this question. Only a small minority consider this option both ideally and in practice to be a good idea.

Finally answer 7 ("Try to find out what the executive's real problem is. It may turn out to have nothing to do with you and rather be connected to personal problems. In which case, you won't feel that you are incompetent at your job") was interesting insofar as nobody considered this to be an extremely bad idea. People were generally comfortable with answer 7, and while there were some who took a neutral stance on the whole this idea was received positively ideally and in practice.

The actual responses of the novices are not of direct interest to us. We are firstly interested to see if the novices respond like experts, and if not, what is it that the experts do that is different. Scenario 12 used in this example has been developed from one of the case studies recorded in [1]. It is interesting to note that option 1 was in fact what the innovator historically chose, though he comments that this option was not very innovative. Instead he recommends option 3 as the most innovative option. This is very interesting because our novices revealed a strong tendency toward intellectual property theft being a bad idea ethically, but starting ones own internet company and bankrupting the competition being a good one in practice. Clearly our novices and our expert have very different views.

Remember that an important part of our research using the inventory was identifying the novelty generation stage a given scenario was at. In the case of Scenario 12, our management students were somewhat divided with regard to the Scenario's innovation development stage. Five students felt the scenario was focusing on *idea generation*, with one of these believing the scenario was concerned with *opportunity recognition* at the same time. The majority of novices (10 out of 23) felt scenario 12 was about *opportunity recognition*. Two out of 23 felt the Scenario was dealing with the *development* stage. And finally 5 students felt the scenario was dealing with the *realisation* stage of innovation.

5 Conclusion and Future Work

What remains to be done next is to extend the results to examine the remaining 11 scenarios with their respective answer options, and then to perform comparisons with that of recognised innovators. More elaborate data analysis techniques such as our use of Formal Concept Analysis [6] should permit us to achieve finer granularity of result analysis than would otherwise be the case with purely statistical approaches.

Most importantly we need to compare the results we have so far with those gained from recognised innovators. A first step in that direction has seen us contact people such as Professor Gordon Bell [1] after whom 'Bells Law of Computer Classes' is named, who was happy to validate our inventory. The next step will be to find other similarly talented individuals who will be identified through innovation awards and ICT organisations specialising in innovative ideas. We seek individuals who are successful both in a technical as well as an entrepreneurial sense.

The benefits of our approach will be best realised in the HR domain. Once we have developed and validated our innovation inventory, we intend to adapt and extend the tool to allow the scenarios to be randomly assigned to potential and existing employees so that it can be used to identify individuals, and to what extent, they behave similarly to the identified innovators. We will need to devise various algorithms to determine acceptable ranges of behaviour and incorporate the use of weightings to allow some scenarios to be more or less important in generating a score. For personnel selection, the goal would be to provide an innovation index/score ranking applicants to assist with the selection process. The tool may be extended to allow other details regarding other selection criteria to be included to make the process more streamlined.

For training purposes, algorithms will be developed which will provide scores indicating what knowledge is currently lacking in the individual and to propose a training programme for the individual. To achieve this goal we will need to refer to and incorporate other research in the psychology, training and recruitment literature.

We intend to compare our approach to the key psychometric approaches offered for innovation testing. We propose to administer techniques such as MBTI, KAI or other psychology-based techniques in order to correlate our findings with these other approaches and to validate the NGM. For instance, we will test whether certain personality traits and characteristics or motivations correspond to the phases in the NGM.

It can be argued that knowledge only exists when it is inside someone's head. When it comes to tacit knowledge we encounter even greater objections to attempts to capture or measure it as by definition such knowledge is implicit, unspoken and even unspeakable. In seeking to measure and capture innovation type knowledge, we are stepping into even more unchartered and cloudy waters. By building on findings from management and psychology based research, we are hoping to shed light on the nature of innovation knowledge. However, we want to move beyond the debate to look at the behaviour patterns that can be identified in the past successes of innovators and extrapolate from that what it means to be innovative and who has the potential to be so.

References

[1] Bell, G., McNamara, J.F. (1991) *McHigh-Tech ventures : the guide for entrepreneurial success* Perseus Books Publishing L.L.C. New York U.S.A.

[2] Busch, P., Richards, D., (2003) "Building and Utilising an IT Tacit Knowledge Inventory" *Proceedings 14th Australasian Conference on Information Systems (ACIS2003)* November 26-28 Perth Australia.

[3] Busch, P. Richards, D. (2004) "Acquisition of articulable tacit knowledge" *Proceedings of the Pacific Knowledge Acquisition Workshop (PKAW'04)*, in conjunction with *The 8th Pac.Rim Int.l Conf. on AI*, Aug 9-13, 2004, Auckland, NZ, :87-101.

[4] Busch, P., Richards, D., (2005) "An Approach to Understand, Capture and Nurture Creativity and Innovation Knowledge" *Proc. 15th Australasian Conference on Information Systems (ACIS2005)* Nov 30-Dec 2nd, Sydney, Australia.

[5] Fenwick, (2003) Innovation: examining workplace learning in new enterprises *Journal of Workplace Learning* 15(3):123-132.

[6] Ganter, R., Wille, R., (1999) *Formal concept analysis: Mathematical foundations* Springer-Verlag Berlin Germany.

[7] Gough, H. (1981) "Studies of the Myers-Briggs Type Indicator in a Personality Assessment Research Institute" *Fourth National Conference on the Myers-Briggs Type Indicator,* Stanford University, July 1981, CA.

[8] Kirton, M. (2001) "Adaptors and Innovators: why new initiatives get blocked" J. Henry (ed.) *Creative Management* 2nd Edition, Cromwell Press Ltd, London, :169-180.

[9] Leonard, D., Sensiper, S., (1998) "The role of tacit knowledge in group innovation" *California Management Review* Berkeley; Spring 40(3) (electronic).

[10] Schweizer, T.S. (2004) *An Individual Psychology of Novelty-Seeking, Creativity and Innovation* ERIM Ph.D. Series, 48.

[11] Sternberg, R., Wagner, R., Williams, W., Horvath, J., (1995) "Testing common sense" *American psychologist* 50(11) :912-927.

Author Index

Lecture Notes in Artificial Intelligence (LNAI)

Vol. 4130: U. Furbach, N. Shankar (Eds.), Automated Reasoning. XV, 680 pages. 2006.

Vol. 4120: J. Calmet, T. Ida, D. Wang (Eds.), Artificial Intelligence and Symbolic Computation. XIII, 269 pages. 2006.

Vol. 4118: Z. Despotovic, S. Joseph, C. Sartori (Eds.), Agents and Peer-to-Peer Computing. XIV, 173 pages. 2006.

Vol. 4114: D.-S. Huang, K. Li, G.W. Irwin (Eds.), Computational Intelligence, Part II. XXVII, 1337 pages. 2006.

Vol. 4108: J.M. Borwein, W.M. Farmer (Eds.), Mathematical Knowledge Management. VIII, 295 pages. 2006.

Vol. 4106: T.R. Roth-Berghofer, M.H. Göker, H.A. Güvenir (Eds.), Advances in Case-Based Reasoning. XIV, 566 pages. 2006.

Vol. 4099: Q. Yang, G. Webb (Eds.), PRICAI 2006: Trends in Artificial Intelligence. XXVIII, 1263 pages. 2006.

Vol. 4095: S. Nolfi, G. Baldassarre, R. Calabretta, J.C.T. Hallam, D. Marocco, J.-A. Meyer, O. Miglino, D. Parisi (Eds.), From Animals to Animats 9. XV, 869 pages. 2006.

Vol. 4093: X. Li, O.R. Zaïane, Z. Li (Eds.), Advanced Data Mining and Applications. XXI, 1110 pages. 2006.

Vol. 4092: J. Lang, F. Lin, J. Wang (Eds.), Knowledge Science, Engineering and Management. XV, 664 pages. 2006.

Vol. 4088: Z.-Z. Shi, R. Sadananda (Eds.), Agent Computing and Multi-Agent Systems. XVII, 827 pages. 2006.

Vol. 4087: F. Schwenker, S. Marinai (Eds.), Artificial Neural Networks in Pattern Recognition. IX, 299 pages. 2006.

Vol. 4068: H. Schärfe, P. Hitzler, P. Øhrstrøm (Eds.), Conceptual Structures: Inspiration and Application. XI, 455 pages. 2006.

Vol. 4065: P. Perner (Ed.), Advances in Data Mining. XI, 592 pages. 2006.

Vol. 4062: G.-Y. Wang, J.F. Peters, A. Skowron, Y. Yao (Eds.), Rough Sets and Knowledge Technology. XX, 810 pages. 2006.

Vol. 4049: S. Parsons, N. Maudet, P. Moraitis, I. Rahwan (Eds.), Argumentation in Multi-Agent Systems. XIV, 313 pages. 2006.

Vol. 4048: L. Goble, J.-J.C.. Meyer (Eds.), Deontic Logic and Artificial Normative Systems. X, 273 pages. 2006.

Vol. 4045: D. Barker-Plummer, R. Cox, N. Swoboda (Eds.), Diagrammatic Representation and Inference. XII, 301 pages. 2006.

Vol. 4031: M. Ali, R. Dapoigny (Eds.), Advances in Applied Artificial Intelligence. XXIII, 1353 pages. 2006.

Vol. 4029: L. Rutkowski, R. Tadeusiewicz, L.A. Zadeh, J.M. Zurada (Eds.), Artificial Intelligence and Soft Computing – ICAISC 2006. XXI, 1235 pages. 2006.

Vol. 4027: H.L. Larsen, G. Pasi, D. Ortiz-Arroyo, T. Andreasen, H. Christiansen (Eds.), Flexible Query Answering Systems. XVIII, 714 pages. 2006.

Vol. 4021: E. André, L. Dybkjær, W. Minker, H. Neumann, M. Weber (Eds.), Perception and Interactive Technologies. XI, 217 pages. 2006.

Vol. 4020: A. Bredenfeld, A. Jacoff, I. Noda, Y. Takahashi (Eds.), RoboCup 2005: Robot Soccer World Cup IX. XVII, 727 pages. 2006.

Vol. 4013: L. Lamontagne, M. Marchand (Eds.), Advances in Artificial Intelligence. XIII, 564 pages. 2006.

Vol. 4012: T. Washio, A. Sakurai, K. Nakajima, H. Takeda, S. Tojo, M. Yokoo (Eds.), New Frontiers in Artificial Intelligence. XIII, 484 pages. 2006.

Vol. 4008: J.C. Augusto, C.D. Nugent (Eds.), Designing Smart Homes. XI, 183 pages. 2006.

Vol. 4005: G. Lugosi, H.U. Simon (Eds.), Learning Theory. XI, 656 pages. 2006.

Vol. 4002: A. Yli-Jyrä, L. Karttunen, J. Karhumäki (Eds.), Finite-State Methods and Natural Language Processing. XIV, 312 pages. 2006.

Vol. 3978: B. Hnich, M. Carlsson, F. Fages, F. Rossi (Eds.), Recent Advances in Constraints. VIII, 179 pages. 2006.

Vol. 3963: O. Dikenelli, M.-P. Gleizes, A. Ricci (Eds.), Engineering Societies in the Agents World VI. XII, 303 pages. 2006.

Vol. 3960: R. Vieira, P. Quaresma, M.d.G.V. Nunes, N.J. Mamede, C. Oliveira, M.C. Dias (Eds.), Computational Processing of the Portuguese Language. XII, 274 pages. 2006.

Vol. 3955: G. Antoniou, G. Potamias, C. Spyropoulos, D. Plexousakis (Eds.), Advances in Artificial Intelligence. XVII, 611 pages. 2006.

Vol. 3949: F.A. Savacı (Ed.), Artificial Intelligence and Neural Networks. IX, 227 pages. 2006.

Vol. 3946: T.R. Roth-Berghofer, S. Schulz, D.B. Leake (Eds.), Modeling and Retrieval of Context. XI, 149 pages. 2006.

Vol. 3944: J. Quiñonero-Candela, I. Dagan, B. Magnini, F. d'Alché-Buc (Eds.), Machine Learning Challenges. XIII, 462 pages. 2006.

Vol. 3937: H. La Poutré, N.M. Sadeh, S. Janson (Eds.), Agent-Mediated Electronic Commerce. X, 227 pages. 2006.

Vol. 3932: B. Mobasher, O. Nasraoui, B. Liu, B. Masand (Eds.), Advances in Web Mining and Web Usage Analysis. X, 189 pages. 2006.

Vol. 3930: D.S. Yeung, Z.-Q. Liu, X.-Z. Wang, H. Yan (Eds.), Advances in Machine Learning and Cybernetics. XXI, 1110 pages. 2006.

Vol. 3918: W.-K. Ng, M. Kitsuregawa, J. Li, K. Chang (Eds.), Advances in Knowledge Discovery and Data Mining. XXIV, 879 pages. 2006.

Vol. 3913: O. Boissier, J. Padget, V. Dignum, G. Lindemann, E. Matson, S. Ossowski, J.S. Sichman, J. Vázquez-Salceda (Eds.), Coordination, Organizations, Institutions, and Norms in Multi-Agent Systems. XII, 259 pages. 2006.

Vol. 3910: S.A. Brueckner, G.D.M. Serugendo, D. Hales, F. Zambonelli (Eds.), Engineering Self-Organising Systems. XII, 245 pages. 2006.